Where Corals Lie

Where Corals Lie
A Natural and Cultural History

J. Malcolm Shick

REAKTION BOOKS

For Jean

*and in memory of
Charlotte Preston Mangum (1938–1998)
Teacher, Mentor and Friend*

Published by Reaktion Books Ltd
Unit 32, Waterside
44–48 Wharf Road
London N1 7UX, UK
www.reaktionbooks.co.uk

First published 2018
Copyright © J. Malcolm Shick 2018

The Coral Reef, composed by Horace Keats (1895–1945), text by John Wheeler.
Permission to reproduce extract given by Wirripang Pty Ltd, Australia.

All rights reserved

No part of this publication may be reproduced, stored in a retrieval system,
or transmitted, in any form or by any means, electronic, mechanical, photocopying,
recording or otherwise, without the prior permission of the publishers

Printed and bound in China
by 1010 Printing International Ltd

A catalogue record for this book is available
from the British Library

ISBN 978 1 78023 934 7

Contents

Prelude: Where Corals Lie *7*

1 Defining Coral *11*

2 On the Nature of Corals *43*

3 The Mythos, Menace and Melancholy of Corals *77*

4 Conjuring Corals *115*

5 Coral as Commodity *153*

6 Coral Construction *187*

7 A New Age of Corals *219*

Coda: What Lies Ahead? *261*

Appendix: Maps Showing Locations Mentioned in the Text *266*
References *270*
Select Bibliography *282*
Glossary *284*
Acknowledgements *290*
Photo Acknowledgements *292*
Index *294*

Prelude: Where Corals Lie

> Coral . . . presented a problem in the eighteenth century . . . since with the development of natural history it finally became necessary to say whether it was animal, vegetable or mineral. Perversely, it exhibited characteristics of all three.
> James Hamilton-Paterson, *The Great Deep: The Sea and its Thresholds* (1992)

In Richard Garnett's mid-Victorian poem that lends its title to this book, the poet is allured to drowning in 'the land where corals lie'. This trope continued a long tradition wherein the reefs built by stony corals were associated with human passing, pathos and petrifaction, and jagged reefs were rightly feared as lethal navigational hazards in maritime expeditionary narratives by the likes of James Cook and Jules Dumont d'Urville, and in novels they inspired. The distantly related precious red corals were born of death – the decapitation of the Gorgon Medusa by Perseus and the supernatural transformation of bloodied seaweeds, or from the flowing blood of the demon king Bali, crushed by Vishnu.

Owing to their rough arborescence, for centuries colonial corals were lyrically described as submarine plants of wonderfully diverse botanical construction and coloration, or perhaps the mineral productions of plants. For others they were stony concretions precipitated from seawater, without any biological mediation. This mineral nature was especially obvious in those coral reefs that wrecked ships. The 'flowers' on coral trees seemed also to have animal attributes, thus completing the tripartition that kept calcareous corals mysterious and equivocal – they were variously called 'lithophyte' (stony plant); 'madrepore' (mother of stone, from Greek *poros*, not pore, or alternatively 'porous mother', from Latin *porus*); 'millepore' (thousand pores [*porus*]); and even 'zoophyte' (animal–plant), a term applied to sundry ill-fathomed invertebrates. Such ambiguity lent itself to manifold interpretations, including mythological and magical imaginings, as well as philosophical musings on the very nature of life (illus. 2, 3 and 4).

Speaking through a chorus of sea beasts in *The Temptation of Saint Anthony*, Gustave Flaubert harked back to the situation before the nature of corals was known:

> Plants are no longer distinguishable from animals. Polyparies [the skeletal framework of coral colonies], which have the air of sycamores, bear arms on their branches . . . And then the plants become confused with stones. Stones look like brains, stalactites like breasts, *fleurs de fer* like tapestries ornate with figures.[1]

Thus, chimeric corals confounded classification and were what the thoughtful (or susceptible) viewer chose to make of them, causing the art historian Barbara Maria Stafford to muse, 'What do you do with beings that are neither one thing nor the other?'[2]

The animality of the coral polyp was confirmed by the mid-1700s, when modern science was emerging and colonialism took

1 Pandora Reef, April 2017, 2–3 m (7–10 ft) depth, photographed by Eric Matson. This coral community contains both healthy and bleached corals in the aftermath of unprecedented mass bleaching in two successive years on the Great Barrier Reef. Corals that died earlier are overgrown by algae.

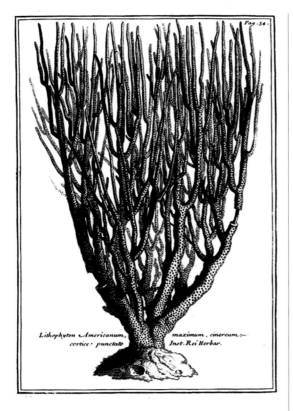

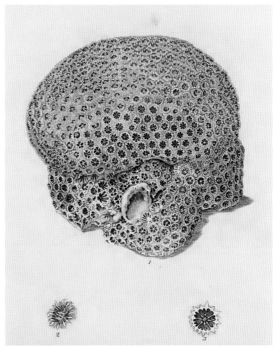

2, 3, 4 Manifestations of corals (clockwise from left) as lithophyte (p. 34 in [Joseph Pitton de] Tournefort, Académie royale des sciences *(1700)); madrepore (plate 55 in John Ellis,* The Natural History of Many Curious and Uncommon Zoophytes *(1786)); and millepore (p. 103 in James D. Dana,* Corals and Coral Islands *(1872)). Opposite page: 5 In 1880–81 Gustave Moreau painted the Nereid Galatea (*La Galatée*) hiding from Polyphemus in a grotto, surrounded by taxonomically diverse zoophytes such as red corals, sea anemones and sea lilies (crinoid echinoderms).*

naturalists to lands and archipelagos where coral reefs lay, and where scientists linked biology to geology. Soon afterward, Linnaeus placed stony corals taxonomically in the ancient Lithophyta and precious corals among the Zoophyta, amid a grab bag of poorly known invertebrates. George Johnston, at the opening to his second edition of *A History of the British Zoophytes* in 1847, reminded readers that the compound word 'zoophyte' originally 'designated a miscellaneous class of beings, which were believed to occupy a space between the animal and vegetable kingdoms, and where the characteristics of the subjects of each kingdom met and were intermingled' (illus. 5).

Another half-century after Johnston, corals were found to live in 'reciprocal accommodation' with unicellular algae inside their own cells, a solar-powered symbiosis that allowed coral reefs to thrive in the clear blue desert of tropical oceans depauperate in planktonic prey. This symbiosis (a literal 'zoophyte', if you will) has been a focus of coral study ever since.

Growing familiarity with coral reefs and idyllic atolls in the late nineteenth century and throughout the twentieth changed our perception of them again. Henri Matisse's sojourn in a Tahitian lagoon in 1930 forever affected his art. The aesthetic allure of corals and reefs also captivates receptive scientists who study them in nature and draws a wider public attracted by tropical beaches and lagoons teeming with colourful fishes, or by a scuba-diving holiday in a submarine wonderland that increasingly features in the portfolios of the best nature photographers. Today, reef-building corals have become 'photogenic doe-eyed invertebrates'[3] and their creations (the largest biological structures on earth, visible from space) are seen as 'rainforests of the sea',[4] greenly symbolic of fragile marine biodiversity.

With the loss of the algal partners in temperature-driven events of mass coral bleaching, the stark bone-whiteness of bleached and moribund coral colonies contrasts shockingly with the rainbow palette that inspired Matisse. It would be a further irony were worsening ocean acidification to corrode the persistent cultural symbol of coral reefs. Despite their massive solidity, increasingly corals are a poster child at risk from, and a canary in the coal mine warning of, global climate change. From dangerous to endangered, ambiguous to emblematic, deathtrap to paradise to boneyard. What has transformed them anew? To answer this question, and as an object lesson of what is at stake in the coming decades, we need to know where corals lie in our social history, including the arts, commerce, geopolitics, philosophy, science and the imagination.

1 Defining Coral

The term 'coral' is neither scientific nor precise.
Daphne G. Fautin and Robert W. Buddemeier, *Encyclopedia of Islands* (2009)

To the ancients, coral was exemplified by what Theophrastus (*c.* 372–*c.* 287 BCE), Aristotle's student, collaborator and successor at the Lyceum, saw as κουράλιον (*kouralion*), a stone similar to hematite in its rich redness but which equivocally grew like a root in the sea – what we today know as *Corallium rubrum*, the precious red coral of the Mediterranean (illus. 6). The etymology of 'coral' is uncertain, but the Greek *kouralion* and its Latin derivative *corallium* may have Semitic origins in the ancient Hebrew *goral* or Arabic *garal*, for pebble or small stone. Others saw the derivation of arborescent corals being conceptually rooted as a stag's branching antlers (Greek *kéras*; Latin *cornu*) and later linked to the reddish *heart*wood (Latin *cor*) of an oak.[1]

Precious coral's shrub-like form and seeming lack of sensation or irritability (one of Aristotle's defining criteria for animals) led the Greek physician Pedanius Dioscorides (*c.* 40–*c.* 90 CE), serving in Nero's army, and Roman naturalist Pliny the Elder (23–79 CE) to consider red coral and its relatives as marine plants. For others, its calcareous red skeleton placed coral among non-living minerals, and in common parlance for centuries and still today in many cultures, it is often that polished red skeleton of calcium carbonate ($CaCO_3$) that has been called 'coral'. When expeditions sailed beyond the Mediterranean, they discovered the much greater diversity of corals, especially those stony forms that build the tropical reefs that capture the imagination and headlines today.

Evolution of Corals, and the Coral of Life

In formal biological classification (taxonomy), all corals are members of the phylum Cnidaria,[2] named for its unique stinging cnidae, which include nematocysts – explosive capsular inclusions in cells that extrude an envenomed barbed tube or thread used, especially, in defence and to capture prey. Many cnidarians, including their simplest representatives, do not have calcified skeletons, so definitive fossil evidence of the phylum's ancient origin is exceedingly rare, but some specimens date to the earliest Ediacaran period, 635 million years ago (MYA).[3] Using the molecular clock (the rate of accumulation of genetic mutations in a group of organisms, or taxon) to deduce the history of the taxon and when it diverged from others places the origin of the Cnidaria even earlier, between 819 and 686 MYA, in the Cryogenian.

Thus cnidarians fall within one of the oldest animal *embranchements* of the tree of life, the Radiata, placed there by the naturalists Jean-Baptiste Lamarck and Georges Cuvier because of their symmetry, which radiates outward from a central axis, with the mouth at the centre of the oral disc, itself surrounded by retractable prey-capturing tentacles laden with nettling cnidae. Cnidarians have only two thin layers of tissue (outer epidermis and inner gastrodermis), which sandwich a jelly-like mesoglea between them. These flat, laminated, two-dimensional layers (imagine a thin sheet of cardboard) are folded, as in origami, to yield three-dimensional animals.[4] The bodies of cnidarians take two forms: polyps (so-called for a resemblance to a tiny

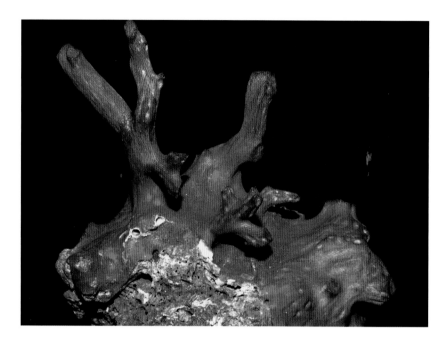

6 Corallium rubrum. *To the ancients, this was coral. A walker on the Mediterranean strand would have seen such a specimen tossed up after a storm, the skeleton denuded of living tissue and slightly polished by tumbling in the waves but still attached to its substratum.*

octopus or *poulpe*, with their circular array of tentacles around the mouth) and medusae, or 'jellyfishes' (so-called because they have much mesoglea and can swim). Some cnidarian life cycles include both forms at successive stages. Corals and other anthozoans exist always as polyps, sessile and attached to the substratum for support.

Cells of the epidermis secrete the multifarious external skeletons (exoskeletons) of 'corals'. Gastrodermal cells are the sites of digestion and also may house living, photo-synthesizing single-celled algae known as zooxanthellae in a nutritional symbiosis. The gastrodermis lines an inner digestive cavity whose single opening serves both as mouth and anus. The cavity is radially divided lengthwise by fleshy mesenteries that reinforce the body wall and are involved in digesting prey and also contain the gonads (illus. 7).

The soft bodies of corals secrete skeletons having varying degrees of calcification and solidity, which help to define different coral groups or taxa. Corals may be small and solitary, consisting of a single polyp, or they may be colonial, consisting of many polyps that remain connected to each other after asexual, vegetative proliferation known

7 Hinrich Nitsche, Corallium rubrum, *Leuckart Chart, Series I, Chart 1: Coelenterata; Anthozoa; Octactinaria, 1877. A colony of the octocoral* Corallium rubrum *comprising multiple polyps, some withdrawn into the coenenchyme (fleshy tissue) that surrounds the calcareous axis of the skeleton, and others expanded, showing the eight pinnate tentacles radially arranged around the central mouth. Longitudinal (top left) and transverse (top centre) sections through an individual polyp show the two tissue layers, outer ectoderm (shown in blue) and inner gastrodermis (yellow), separated by a gelatinous mesoglea (pink). The transverse section shows the radial arrangement of the mesenteries in the digestive cavity. The longitudinal section shows the gonads (here, ovaries) in the mesenteries. The fertilized egg develops into a planula larva (centre left) that settles and becomes a primary polyp, which buds to produce additional polyps that form the colony (bottom left).*

as budding. Coral colonies can grow very large, measuring in metres.

The broad evolutionary relationships of corals are most easily grasped graphically, in a phylogenetic tree, or 'tree of life'. Most corals are colonial associations of many interconnected polyps that inhabit a communal skeleton, and early on Charles Darwin (in his notebook of 1837, the year after he returned from his global circumnavigation aboard HMS *Beagle*) mused that 'The tree of life should perhaps be called the coral of life.' This depiction occurred to him particularly because the calcareous remains in the lower branches (where the polyps had died as the colony grew upward) were as a fossil record of extinct species, whereas the upper branches containing live polyps represented extant species whose positions manifested their relationship to each other (illus. 8).

Most corals belong to the class Anthozoa ('flower animals'). Such verbal equivocation characterizes the conundrum of coral classification. The earliest non-mineralized anthozoan fossils may be from the lowest Cambrian, around 540 MYA.[5] Molecular clock estimates push the origin of the Anthozoa back to 684 MYA. The exclusively marine class Anthozoa comprises those seemingly floral yet irritable beings – solitary sea anemones, diverse solitary and colonial corals, and others.

An entirely different class of Cnidaria, the Hydrozoa, includes the fire corals (so named for their intensely stinging nematocysts) in the genus *Millepora* and the brilliantly coloured lace corals, whose heavily calcified colonial skeletons evolved

8 The 'coral of life', showing the major taxonomic groups of 'corals' mentioned in the text, and their evolutionary relationships. The Anthozoa and the Medusozoa diverged early, and the former class Scyphozoa has been recognized as a lineage of related groups of 'jellies'. Classification within the Octocorallia remains problematic because various morphologies loosely known as 'soft corals', sea fans and sea whips are distributed throughout the lineages of octocorals. Original drawing by Ryan Cowan.

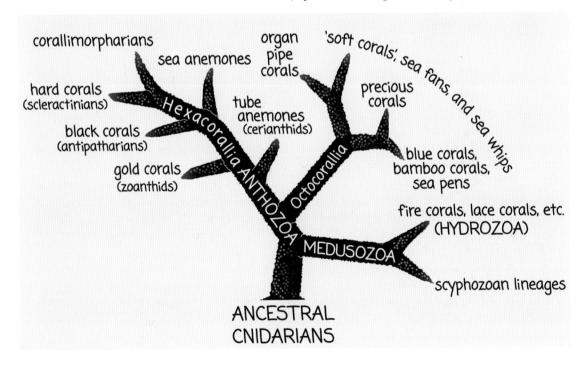

independently of anthozoan skeletons. Unlike anthozoans, many hydrozoans include a medusa in their life cycle, which unites them with the familiar 'jellyfishes' of the several classes in the lineage that includes the Scyphozoa among the broader Medusozoa. The Medusozoa split from the Anthozoa at least 543 MYA (based on the fossil record) or up to 642 MYA (the molecular clock estimate).

Mediterranean red coral – *Corallium rubrum* – was known for millennia but not given its current descriptive Latin binomial until Lamarck did so in 1816. *C. rubrum* is but one of some two dozen 'precious corals' in the family Coralliidae. The Coralliidae falls in one taxonomic order (Alcyonacea) in one subclass (Octocorallia, named for the eight radially arranged pinnate tentacles) of the wider class Anthozoa. The earliest octocorals split from the other subclass of anthozoans, the Hexacorallia (see below), no later than 540 MYA but possibly earlier, in the Ediacaran period. The colonial sea pens are an ancient, soft-bodied group (order Pennatulacea) dating at least to the Cambrian, as represented by a fossil in the Burgess Shale (505 MYA).

Red corals take dissolved calcium (Ca^{2+}) and carbonate (CO_3^{2-}) ions from seawater and precipitate solid crystals of calcium carbonate ($CaCO_3$) as the mineral calcite of their skeletons. Other less calcified alcyonaceans sometimes loosely referred to as 'corals' include the sea whips and sea fans, all informally associated as 'gorgonians'. In red corals and other precious octocorals, the polyps themselves do not secrete the skeleton. A layer of skeletogenic cells produces a solid, unitary axial core of $CaCO_3$, surrounded by the coenenchyme (a communal tissue where each polyp can withdraw into its individual cup-like calyx). Other cells scattered throughout the coenenchyme secrete small calcareous sclerites (illus. 9). Colonies of sea fans and sea whips, on the other hand, support themselves with a proteinaceous axis of gorgonin (which includes

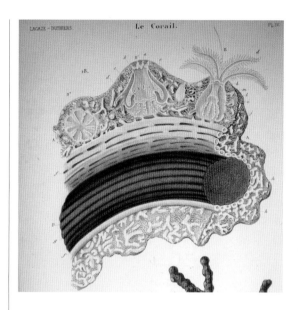

9 H[enri] Lacaze-Duthiers, Histoire naturelle du corail *(1864)*, plate IV, detail. *The interior of a colony of* Corallium rubrum, *a representative of the precious corals in which the calcareous skeleton is a central axis that is biochemically tinted in shades of red. The individual polyps are interconnected via tubes within the fleshy mass of the coenenchyme.*

a chemically tanned collagen, the protein found in the connective tissue of many groups of animals) and similarly embed calcareous sclerites in the soft coenenchyme (illus. 10).

Other modern alcyonaceans, which began to diversify during the Triassic recovery from the worldwide end-Permian extinction about 252 MYA, include several different groups of 'soft corals', including *Alcyonium digitatum* (dead man's fingers), that lack a central axis and are only loosely supported by scattered calcareous sclerites that consolidate the coenenchyme (illus. 11 and 12). Many alcyonaceans intergrade (pass into another form) in their morphological characters, some of which arose independently in the different groups, complicating the taxonomy of the Octocorallia. Their phylogeny (evolutionary relationships) is increasingly based on molecular genetic analyses.

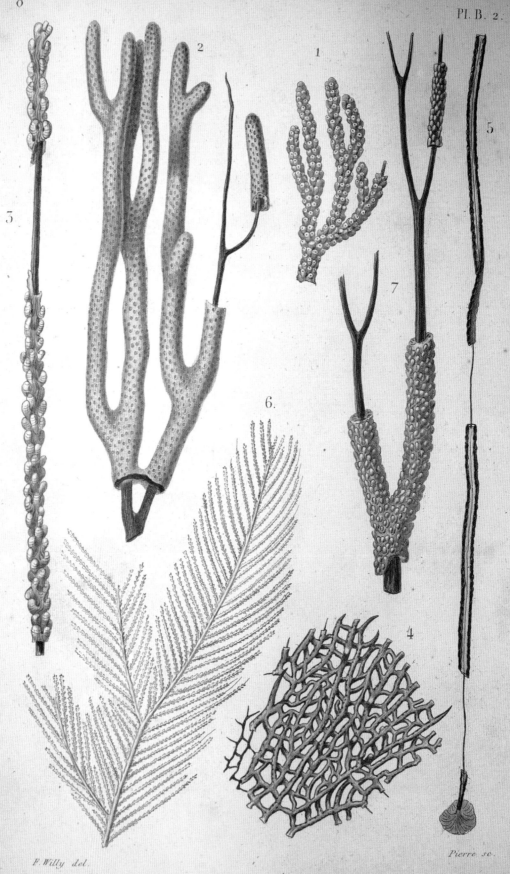

Alcyonaires.

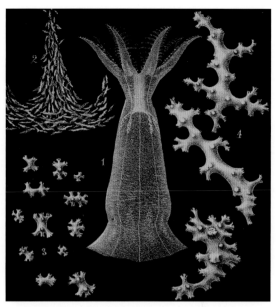

11 and 12 *A colony of the 'soft coral'* Alcyonium digitatum, *the 'dead man's fingers' (left; George Johnston,* A History of the British Zoophytes *(1838), plate XXVI); an individual polyp and the calcareous sclerites that support the tissues of the colony (right; Philip Henry Gosse,* A Naturalist's Rambles on the Devonshire Coast *(1853), plate III).*

The uncertainties of relationships among taxa of octocorals, especially soft, whip-like and fan-like forms, are apparent in illus. 8.

Still another family of alcyonaceans includes the semi-precious 'bamboo corals', which have a more porous skeleton (with long sections consisting mostly of gorgonin, punctuated by nodes of calcium carbonate, giving the branches the look of bamboo) (illus. 13). *Heliopora coerulea* is in a different order, the Helioporacea or blue corals, important members of coral reef communities, named for their massive blue skeleton of the calcareous mineral aragonite, exceptional among octocorals.

The other anthozoan subclass, the Hexacorallia (whose members have multiples of six radially arranged tentacles), in addition to containing the familiar skeleton-less sea anemones (order Actiniaria), the tube anemones or cerianthids (order Ceriantharia) and the less familiar order Corallimorpharia, additionally encompasses three orders of corals. The order Antipatharia (which takes its name from its sometime medical use to combat pathological conditions) comprises only the semi-precious, often tree-like black corals known to the ancients, who used them not only medicinally but also

Opposite page: 10 Henri Milne-Edwards, Histoire naturelle des coralliaires ou polypes proprement dits *(1857), vol. I, plate B2. Cutaway views of several sea whips (top) showing the central proteinaceous axis.*

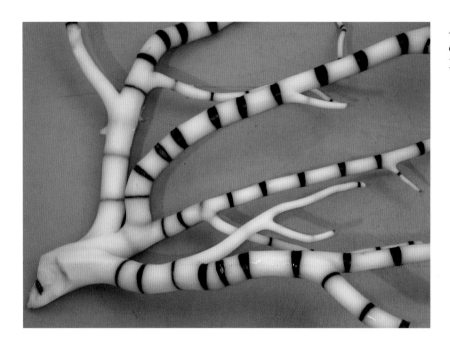

13 The dried skeleton of a bamboo coral, Keratoisis flexibilis.

as divining rods and princely sceptres. Black coral is familiar today as the state gem of Hawaii. Black coral skeleton is not calcareous, but rather a largely proteinaceous complex called antipathin that unlike gorgonin includes not collagen but the polymeric carbohydrate chitin (the stuff of crustacean exoskeletons), and which takes a high polish (illus. 14).

Another ancient order, the Zoanthidea, includes the rare Hawaiian gold coral, which likewise forms a proteinaceous skeleton used in making jewellery. Like many black corals, gold coral inhabits seamounts and other deep-water structures. The extreme longevity of gold (>2,500 years) and some black coral (>4,000 years – the oldest living animals)[6] colonies has made their slow-growing skeletons useful as long-term recorders of oceanic chemistry and productivity (illus. 15).

But the best-known corals, the principal builders of coral reefs, are in the hexacorallian

14 In life, large colonies of antipatharian black corals may have sponges, oysters, echinoderms or other invertebrates growing on them for support, still present on this large dead colony of Antipathes dichotoma *(now* A. griggi*) from a depth of 65 m (213 ft) off Maui, Hawaiian Islands.*

order Scleractinia (illus. 16). They thrive in warm, shallow, sunny tropical and subtropical seas extending to 30° North and South of the equator, most abundantly in the Coral Triangle which has its corners at the Philippines, Malaysia and the Solomon Islands (illus. 17). The polyps in these 'hard corals' precipitate the calcareous mineral

15 The robotic arm of a submersible collecting a Hawaiian gold coral, Kulamanamana haumeaae.

16 The growth-forms of scleractinian corals contribute to the structural habitat diversity of a coral reef. The basic growth-forms are (clockwise from bottom left): encrusting (closely adhering to the substratum); laminar (forming a tier); massive; branching; foliaceous (forming a whorl); columnar; and free-living (not fixed to the substratum). Illustration by Geoff Kelley in J.E.N. Veron, Corals of the World (2000), vol. 1, pp. 56–7.

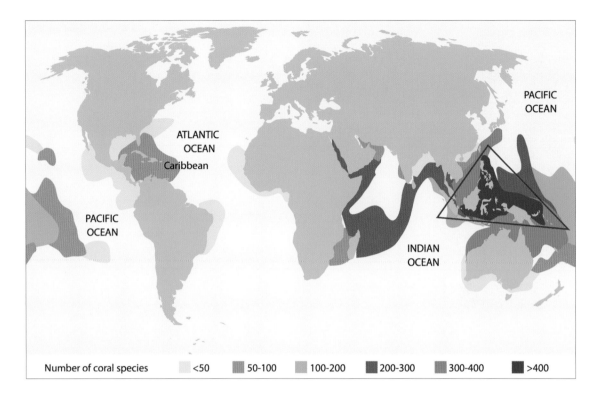

aragonite of their massive skeletons, assisted chemically by their unicellular endosymbiotic algae, the zooxanthellae. Other scleractinians (some harbouring, others lacking zooxanthellae) occur worldwide, from shallow to deep water (where eternal darkness precludes photosynthesis), including at polar latitudes. Although long known to deep-water dredgers, some scleractinians inhabiting cold (but not always deep) water are being rediscovered today in an era of intensified offshore trawling, oil exploration and biodiversity awareness. Many of these corals are solitary, but others form extensive mounds, banks or reefs of reticulated colonies, although not as solidly massive as the cemented reef structures familiar in shallow tropical seas (illus. 18).

Scleractinians are relative newcomers on the coral scene, appearing first as fossils about 237 MYA, during the Triassic recovery following the worldwide extinctions at the end of the Permian period 252 MYA.[7] The Tabulata and the Rugosa,

17 *The global distribution of coral reefs and the diversity of reef-building corals. The Coral Triangle (bounded in red) includes the highest biodiversity of reef corals on earth, some 500 species.*

two prominent earlier taxa of calcitic corals (their fossils dating to the early and mid-Ordovician, 490 and 470 MYA, respectively), were lost along with the vast majority of marine species in the end-Permian mass extinction (illus. 19). Various biogeochemical causes have been invoked for this Great Dying, some of which (ocean acidification and anoxia, or absence of oxygen) have been implicated in more recent marine extinctions and 'reef gaps' in the fossil record and increasingly bode ill for today's coral reefs. Scleractinians arose (possibly more than once, forming different lineages) from soft-bodied survivors resembling sea anemones when seawater conditions permitted the deposition of their aragonitic skeletons. Conversely, corallimorpharians seem to be

18 Philip Henry Gosse, A History of the British Sea-anemones and Corals *(1860)*, plate x. Cold-water (or deep-sea) scleractinian corals include reticulated colonies (Lophelia pertusa, = L. prolifera, *top left)* or solitary cup-corals (Caryophyllia = Cyathina smithii, *top centre,* and Balanophyllia regia, *top right).*

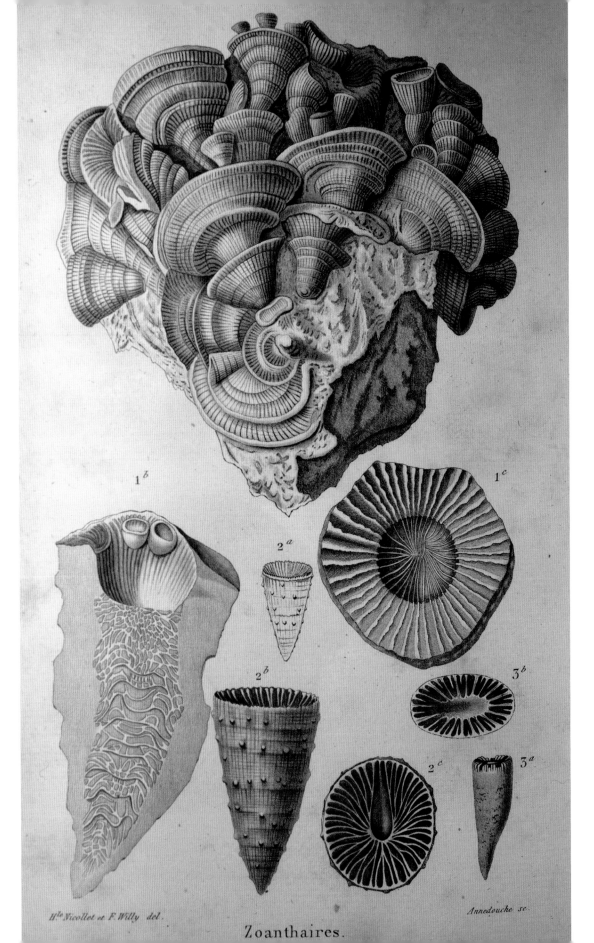

Zoanthaires.

scleractinians that lost their skeleton yet survived during the Cretaceous, when oceanic carbon dioxide (CO_2) and acidity were high. Calcitic precious corals descended from a line of octocorals that survived the end-Permian extinctions.

The Life of Corals

An individual polyp may develop from the sexual union of male and female gametes – sperm and egg. Coral polyps may be either female or male, or hermaphroditic – having gonads of both sexes in a single individual. Eggs and sperm may be released separately, or together in packets from hermaphroditic colonies, for external fertilization. Coral colonies on Australia's Great Barrier Reef engage in synchronized mass spawning several nights after a November full moon in austral springtime (sensationalized in popular accounts as orgiastic group-sex encounters). The timing assures that the intermingled eggs and sperm can join in fertilization. This 'broadcast spawning' of gametes is so predictable that the spectacle is an attraction for tourists and journalists, and researchers can plan on collecting gametes annually for experiments on the corals' reproductive biology. Gametes cloud the water in a pastel submarine blizzard (illus. 20), and the morning after, myriads of developing embryos and dying unfertilized eggs slick the surface in shades of orange and rose (illus. 21).

Although she artistically anticipates the first published description of such an event by more than a century and a half, and relocates the reefs geographically, Rebecca Stott uses the scene as a poetic coda to her novel *The Coral Thief* (2009), set in Paris in 1815:

19 Henri Milne-Edwards, Histoire naturelle des coralliaires ou polypes proprement dits *(1857), vol. III, plate G1. Fossilized skeletons of solitary rugose corals. Notice the bilateral (mirror-image) symmetry of the skeletal septa (bottom right), unlike the radial symmetry seen in most scleractinian and tabulate corals.*

20 and 21 Underwater photograph of colonies of Acropora *spawning (top), and aerial photograph of spawning slicks on the sea surface over patch reefs, each a few tens of metres across (bottom).*

She had seen Red Sea coral spawn, she said. When the sea reaches the right temperature, when they are ripe, when the moon reaches a certain point, just once a year, down there on the coral reefs, the dark waters explode into white smoke clouds. It's like fireworks or seed heads opening, thousands and millions of them, released into the water all at once . . . The fishermen say it's the moon that makes them spawn, she had said, and I said: How can they see the moon? They have no eyes. Perhaps they have other ways of seeing and knowing, she had said.

Anna Thynne (1806–1866), the originator of the Victorian marine aquarium (creating one in Westminster Abbey, where she lived with her husband, the sub-dean of the abbey), saw the synchronous spawning of several captive specimens of the solitary cup coral, *Caryophyllia*: 'One would think they had some curious sympathy with each other, that so many of them should be similarly engaged exactly at the same moment.'[8] On the reef, it is not a curious sympathy, or some mystical way of seeing or knowing, but rather their cryptochromes or melanopsin – proteins sensitive to blue light with which eyeless corals detect the moonlight and set their rhythmic biological clocks – that synchronize their spawning.[9]

In some species fertilization is internal, with the polyp brooding the embryos up to the stage of the tiny planula larva (see illus. 7). Whether developing from external or internal fertilization, the planula may settle quickly or drift for hours, days or weeks in the plankton, providing a dispersal stage for the otherwise sessile coral. When it does settle, it cements itself to the substratum and develops into a primary polyp, which buds to produce additional polyps that build a colony (see illus. 7). For zooxanthellate scleractinians (which harbour zooxanthellae) it is important that the planulae settle in shallow water in the ocean's photic zone, where there is enough light for their algae to photosynthesize.

Even in the sunless deep, azooxanthellate scleractinians and other coral taxa that normally lack symbiotic algae may exist as solitary individuals or form reefs, mounds or bioherms of loosely consolidated calcareous colonies. Suitable hard substratum may be difficult to find, particularly on the vast expanses of deep-sea muds and oozes. Eventually, a wide-ranging opportunistic planula will settle for anything. This includes human artefacts deposited in the deep by shipwreck or other misadventure, or deliberately. The 1850s saw the first laying of undersea telegraph cables. Robert Louis Stevenson wrote of the electrical engineer (and first professor of engineering at the University of Edinburgh) Fleeming Jenkin, who participated in a seagoing expedition to grapple broken cables (damaged by draggers for precious coral) from the seabed between Sardinia and North Africa. Stevenson, working from Jenkin's diary, described an unanticipated aesthetic experience:

> Yesterday the cable was often a lovely sight, coming out of the water one large incrustation of delicate, net-like corals and long, white curling shells. No portion of the dirty black wires was visible; instead we had a garland of soft pink with little scarlet sprays and white enamel intermixed.[10]

Corals' carpeting the cable was possible because, unlike Anna Thynne's solitary cup coral, most scleractinians and alcyonaceans are colonial, and a single founding polyp may produce buds that develop into multiple new polyps and establish a modular colony of thousands of connected clonal inhabitants spreading vegetatively (see illus. 7). This type of asexual proliferation, akin to that in many terrestrial plants, requires a highly plastic development capable of generating new polyps indefinitely, a characteristic of many corals. Such a colony grows indeterminately (that is, potentially infinitely) as each budding of individual polyps yields two or more additional polyps, which each bud in turn. In scleractinians, each new polyp secretes its own calcareous, cup-like corallite, divided radially like spokes of a wheel by septa that correspond to the internal anatomy of the polyp's fleshy mesenteries (illus. 22 and 23). Septa are added in sequential cycles, and both the number of cycles and the morphology of the septa and the corallites (because of their crystalline permanence even after the death of the polyp) historically have been used in coral systematics – the description of species and their evolutionary relationships. Polyps

22 and 23 A colony of the scleractinian hexacoral Stylophora pistillata *is formed by the budding of individual polyps or differentiation of the tissue between them to produce new polyps (top), each of which inhabits a corallite in the calcareous skeleton (right). The hexamerous radial symmetry is evident in the twelve tentacles of the polyps, and the six septa of the corallites. The polyps and their corallites are 1–2 mm in diameter.*

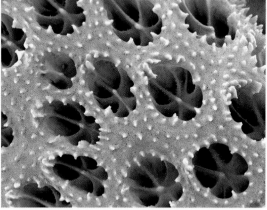

are not always discrete; their oral discs may form a long, sinuous ribbon containing serial mouths, characteristic of meandrine forms such as brain corals (illus. 24).

Scleractinian polyps nestled in their corallites remain connected to each other by the coenosarc, a living tissue that covers the skeleton between the corallites. The colony enlarges its surface area and thickness as the polyps secrete and inhabit new corallites atop the old ones, which they vacate and wall off by thin bulkheads; the polyps are not entombed in the process (as some melancholic Romantic poetry by James Montgomery suggests), and because each budding produces additional polyps that collectively comprise a single clone of genetically identical individuals, they and the colony are potentially immortal – evoking hope, not pathos.

By progressing ever upward and outward, and leaving behind the unoccupied skeleton as the base and bulk of the colony, the polyps remain at the surface of the structure. 'A coral reef, therefore, is a mass of brute matter, living only at its outer surface, and chiefly on its lateral slopes.'[11] This topological distribution is important because the zooxanthellate polyps must be exposed to the surrounding environment to capture food for themselves and nutrients for their algal endosymbionts, which moreover must be exposed to sunlight in order to photosynthesize. The living polyps form a delicate veneer on the surface of the massive skeleton. Like most of us, corals spread themselves thin to make a living (illus. 25).

Joseph Beete Jukes, the geologist and naturalist on the surveying voyage of HMS *Fly* off northern Australia, wrote in his 1847 narrative of seeing 'very large blocks and crags of a porites, twenty feet long and ten feet high, but all one connected mass, without any breaks in its growth'. Cylindrical cores drilled from such massive colonies place the ages of some as greater than half a millennium (illus. 26).

24 A meandrine 'brain coral' on the Great Barrier Reef.

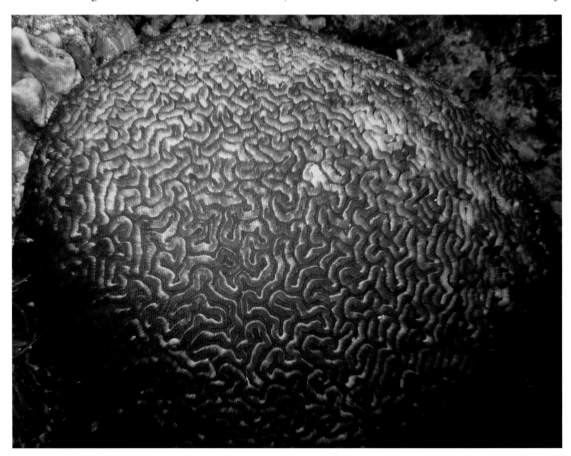

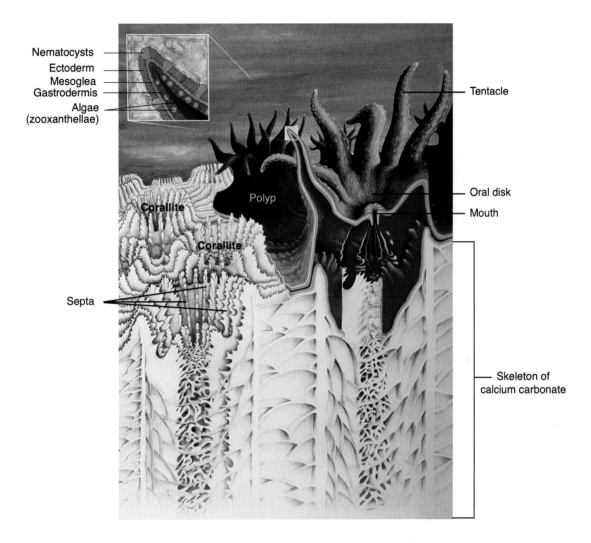

25 *The thin veneer of living coral tissue sits atop the massive skeleton of calcium carbonate secreted by the individual polyps in their corallites. The inset shows the detail of a tentacle with its outer ectoderm containing stinging nematocysts, separated by the gelatinous mesoglea from the inner gastrodermis, cells of which contain endosymbiotic algae (zooxanthellae). The colony grows as the polyps bud to produce additional polyps, which secrete and inhabit new corallites atop the old. Illustration by Geoff Kelley in J.E.N. Veron,* Corals of the World *(2000), vol. I, p. 48.*

Such hale longevity involves minimizing 'the thousand natural shocks that flesh is heir to'. These shocks include oxidative stress evoked in the coral cells by environmental factors such as pollutants and radiation (including sunlight) that cause the production of reactive oxygen species (ROS) toxic to living tissues. The large size of ancient coral colonies involves the ongoing generation of new polyps from portions of old ones or from the coenosarc between polyps, afforded by perennially undifferentiated stem cells that have vast potential to divide and develop into new tissues. The eighteenth-century Swiss

naturalist Charles Bonnet suspected as much and, sparked by his cousin Abraham Trembley's experiments on regeneration in freshwater *Hydra*, posited the existence of ageless 'sleeping embryos' whose 'souls' awakened when a body part was lost and grew to replace it. Such regeneration is the norm in many sea anemones and solitary corals that reproduce by binary fission, with each half regenerating the parts it has lost. Stem cells and regeneration, and defences against oxidative stress, are of intense interest to biomedical scientists.[12]

Molecular oxygen, O_2, is used by the cell in the tiny mitochondria (the cell's energy powerhouses) for the controlled burning in respiration of carbon fuels obtained from digested prey or via the zooxanthellae. Mitochondria also are a major source of the toxic ROS associated with this slow combustion. Antioxidant enzymes and DNA-repair machinery are especially robust in the mitochondria (where the DNA, a vestige of an ancient symbiosis with bacteria, is distinct from that in the cell's nucleus). One result of these active defences is that the accumulation of mutations and genetic change in the mitochondrial DNA of cnidarians – especially several groups of corals and particularly the octocorals – is exceptionally slow. This low rate of mitochondrial genome evolution explains much of the difficulty in discerning phylogenetic relationships among the octocorals, evident in illus. 8.

Living for centuries as do some scleractinians, or for millennia as do gold and black corals, may also involve the ongoing replacement of cells during the vegetative budding and generation of new, young polyps.

26 A massive Porites *colony 3.4 m (11 ft 2 in.) tall on Clerke Reef, northwest Australia, being cored for analysis. The colony started growing in about 1737 and was already 65 years old when Matthew Flinders sailed past on his 1801–2 circumnavigation of Australia.*

This 'continual replacement regimen'[13] may involve pluripotent stem cells (which form a large proportion of the total cells in a cnidarian) in a cycle where 'in the living zoophyte, death and life are going on together, *pari passu*,'[14] with any remains being recycled in the colony. Thus the new polyps are rejuvenated and later can bud again before they reach an age where accumulated oxidative insults and physiological deterioration become lethal.

Not all coral colonies reach the size of those massive *Porites* heads described by Jukes and others. And not all corals are immortal, even absenting natural disaster, pollution, disease or being devoured by a crown-of-thorns sea star, *Acanthaster planci*, a devastating predator of Indo-Pacific corals. In apparently healthy colonies of *Stylophora pistillata*, decreases in sexual reproduction and calcification presaged by three to six months the deaths of individual cabbage-sized colonies.[15] How is it, then, that polyps of some, but not all, corals seem to be able to bud and regrow almost indefinitely so as to approach immortality, at least in the eyes of short-lived humans, and allow colony growth to such large sizes?

After many cycles of cell division, the genome becomes unstable, the cell division cycle is arrested and cells senesce and die. This cellular ageing is accompanied by a progressive shortening of the chromosomal telomeres (expendable regions of repetitive base sequences of DNA) that cap the ends of the chromosomes and protect them from gene rearrangement and degradation. Telomere length in aged humans is reduced by about two-thirds compared to that in newborns. The state of the telomeres thus may play a role in cellular ageing and individual life span. Telomere shortening is offset by the enzyme telomerase, which resynthesizes the particular DNA that forms the end-caps.

All species of corals that have been examined have telomeres and telomerase that vary in their length and activity,[16] both in the gametes or their precursor cells (which must avoid senescence if sexual reproduction is to continue in very large colonies) and in somatic tissues (that is, those other than reproductive cells), including stem cells. The data are scant but lead one to wonder whether long-lived corals that form massive colonies might be among those that have longer telomeres and higher telomerase activity, and are better endowed with stem cells, than those that do not. Unfortunately there appear to be no data for those taxonomically diverse coral Methuselahs that live for hundreds or thousands of years.

A continuous record of the scleractinian colony's growth and environment is provided by chemical and structural proxies in crystalline aragonite. The progressive growth of a massive, unbranched colony can be visualized by taking a thin slice through the skeleton of an entire colony, or slicing lengthwise a cylindrical core removed from a larger colony, and making an X-ray of the sliver, revealing the stacks of corallites and the walls between them. Different density bands in the X-ray represent annual growth increments, akin to tree rings, giving the colony's age and growth rate. Particular isotopes of elements and their ratios, and luminescent regions, in the crystalline skeleton can be used as proxies to infer the seawater temperature and salinity, and incidents of terrestrial runoff carrying sediments and nutrients, when the polyps deposited the skeleton (illus. 27).[17] This is possible because 'the quarry from which they dug their masonry was the limpid wave.'[18] Some fossilized tabulate and rugose corals from the Devonian (~340 MYA) even evince daily growth rings, which surprisingly indicate a year of four hundred days at that time.[19] These corals were not lying about their age, but rather provided early chronometric evidence supporting the astronomical calculation that the terrestrial day has been lengthening (that is, the earth's rotation on its day–night axis has been slowing) and thus the number of days per year has been progressively decreasing over millions of years as the moon receded from the earth.

The age of scleractinian skeletons is also measured by precise thorium–uranium (^{230}Th/^{234}U) dating, after living corals absorb radioactive uranium (but not thorium) from seawater and deposit it in their skeletons. When the polyps die and no new uranium is incorporated, skeletal uranium gradually decays to thorium at a known rate and changes the ^{230}Th/^{234}U ratio, which is used to calculate when the coral died.

The skeletal foundation represents the colony's history writ in stone and chemistry, a record that is increasingly read both by anthropologists studying monumental structures built of coral on Pacific islands and by geochemists to understand the environmental conditions under which the skeleton was laid down, important in predicting the effects on coral reefs of projected changes in global climate and ocean chemistry. Today's polyps rest on their concrete past, but await an uncertain future.

The formation of colonies of modular polyps frees corals from the constraints of individual polyp size and shape: coloniality, integral to reef building, increases the types of habitats that corals may occupy and enhances their chances for long-term evolutionary success by spreading the risk of demise among many polyps each bearing the same genotype. Both branching and massive scleractinians are the 'ecosystem engineers' that form the framework of a coral reef. Other organisms of the reef community contribute to its three-dimensional structure. Some of these provide the calcareous debris of their own skeletons that forms sediments mortaring the interstices of the reef framework and strengthening it. A few pages after calling most of a coral reef 'brute matter', Jukes explained that the 'coral conglomerate' was 'formed of coral and the detritus of coral and shells' overlaying the granitic rock of the mainland.

Octocorals such as sea fans, sea whips and other alcyonaceans baffle and slow the water

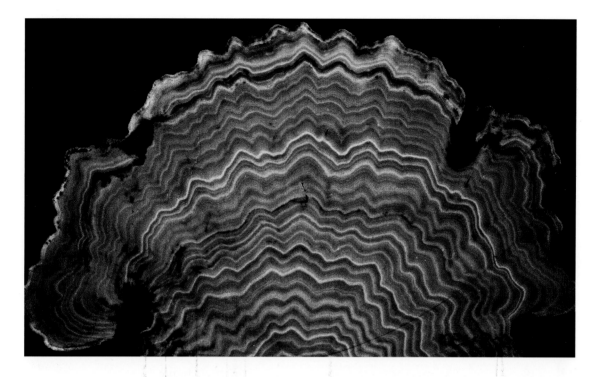

currents moving across the reef, facilitating the deposition of sediments. Crustose coralline red algae help to cement such sediments and larger fragments of coral rubble, cohering the composite. It is these colourful algae that were referred to as 'nullipores' in early descriptions (such as that of Jukes) of the palette of coral reefs (illus. 23). The pink patches provide open space where coral planulae are chemically attracted and eventually settle and grow into new colonies. Deep inside the reef, seawater chemistry and physics, and microbial activity, gradually transform the calcareous debris into sedimentary coral rock in the process of diagenesis.

The converse of reef accretion and diagenesis is bioerosion, in which accumulated calcium carbonate is abraded and solubilized by the tunnelling activities of reef borers (sponges, as well as bivalve molluscs and polychaete worms) or rasping by fishes and sea urchins grazing on algae and on the corals themselves, turning coral skeleton into debris and sand (illus. 29 and 30).

27 Digitally enhanced UV-luminescence photograph of a section through a small Porites *colony (39 cm (1 ft 4 in.) high by 59 cm (1 ft 11 in.) wide) collected in 2004 from the inshore Great Barrier Reef. The luminescent lines are the result of terrestrial organic matter incorporated into the skeleton, and the intensity of each is strongly related to the magnitude of annual freshwater flood events. The oval pit at the upper right was excavated by a boring bivalve mollusc.*

Other invertebrates such as crabs and brittle stars move into the spaces excavated by the borers. Waves and currents, too, erode the reef and carry away sand and sediment, which are deposited as broad flats behind the reef and form idyllic beaches. The smallest particles may eventually be reconsolidated to form sedimentary limestone called oolite, quarried today as a building material. To persist, reefs must accrete calcium carbonate faster than it erodes (illus. 31).

This balance between the accretion and erosion of calcium carbonate is tipping in the favour of erosion because growing global industrialization is escalating the addition of carbon dioxide to the atmosphere and oceans. As we will see in Chapter Seven, when added to seawater, CO_2 forms an acid that underlies 'ocean acidification'. In the process, the normally alkaline seawater does not actually become acidic but rather less alkaline, which is less favourable to the chemistry of calcification and thus diminishes the deposition of calcium carbonate, thereby slowing the growth of coral reefs.

Borers and other bioeroders actually add to the three-dimensional complexity of the reef and provide habitat space for other species, thereby contributing to the high biodiversity of coral reefs – the poet and environmental writer Melanie Challenger's 'gnarled underworlds of many species'. This labyrinthine image figures in fiction as well as ecological treatises; according to Jules Rengade in a novel for young readers, 'All that the Indian Ocean contained in the way of zoophytes, annelids, molluscs and curious fishes met in the mazes and stony passageways of the Coral Sea.'[20] John Steinbeck and Edward Ricketts wrote lyrically of these labyrinths in the narrative portion of *Sea of Cortez* (1941), later published as *The Log from the Sea of Cortez*:

> Clinging to the coral, growing on it, burrowing into it, was a teeming fauna. Every piece of the soft material broken off skittered and pulsed with life – little

28 The coral Agaricia *is growing over the pink crustose coralline alga* Paragoniolithon, *which helps to cement and consolidate the reef.*

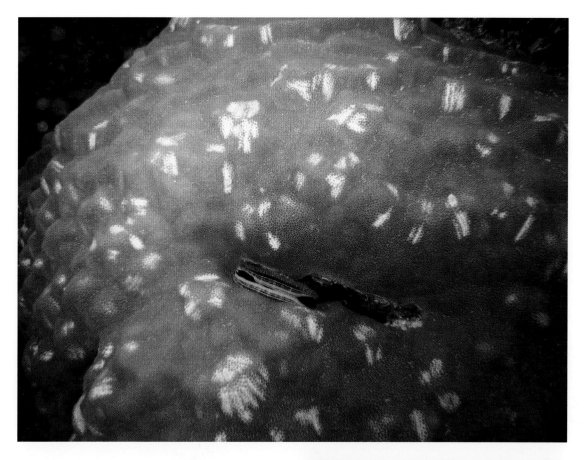

29 and 30 Parrotfishes grazing on algae and corals leave excavation scars where they remove coral tissue and skeleton; the boring bivalve mollusc Pedum *also erodes the coral (top). The bits of calcium carbonate scraped from the coral skeleton pass through the digestive tract of the parrotfish and are deposited as coral sand.*

crabs and worms and snails. One small piece of coral might conceal thirty or forty species, and the colors on the reef were electric . . . Several large pieces of coral were . . . allowed to lie in stale sea water in one of the pans . . . [A]s the water goes stale, the thousands of little roomers which live in the tubes and caves and interstices of the coral come out of hiding and scramble for a new home. (illus. 32)

The 'curious', exotic fishes living in the interstices of coral reefs command high prices for the restaurant and aquarium trades. They can be coaxed from their lairs, dead or alive,

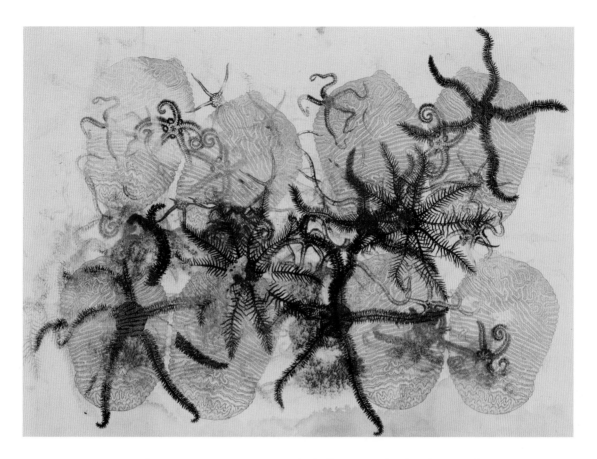

Opposite page: 31 Jörg Schmeisser, Some Fragments on the Beach*, 2011, colour etching; Above: 32 'Clinging to the coral, growing on it, burrowing into it, was a teeming fauna', including brittlestars, as seen in Philip Taaffe's* Sea Stars with Coral II*, 1997, oil pigment on paper.*

by explosives (dynamite, or the reef terrorist's bottles of ammonium nitrate fertilizer cooked with diesel or kerosene) or poisons (cyanide or bleach, toxic to corals as well as to fishes). Both destructive fishing methods are illegal but still used widely and have damaged more than half of the reefs in Southeast Asia.[21]

Deep-water reefs of azooxanthellate scleractinians such as *Lophelia pertusa* living below the depth to which sunlight penetrates lack the extra cementation provided to reefs in shallow water by the photosynthetic crustose coralline red algae. In deep water the reef framework is primarily the product of the polyps per se. The young corals recruit on their living forebears' skeleton, and their growing branches bond to and stabilize the scaffold, where even broken branches may re-fuse to the main structure and not fall away from the living reef into the 20,000-year-old debris at its base. In *L. pertusa*, fusion of the coral colonies compensates for the lack of encrusting coralline algae, enhancing the coral's role as an ecosystem engineer and builder of three-dimensional deep-sea reefs that provide habitat for other invertebrates and fishes and contribute to deep-sea biodiversity (illus. 33).

Being colonial gives corals a competitive edge in the battle for space on reefs, where

33 A Lophelia *reef supports a diverse community that includes fishes such as this cusk, as well as soft corals, sponges and other invertebrates.*

the *Lebensraum* of hard substratum is fiercely contested. Species of solitary corals are scarce on tropical reefs because their individual size is limited and occupation of available hard bottom begins with tiny sexual propagules that develop from planula larvae and can be overtopped or overgrown by established, fast-growing colonial competitors. Such reefs are dominated, indeed characterized, by colonial corals that can quickly monopolize an area by vegetative (asexual) growth and exclude competitors.

In such spatial competition, 'the race is not only to the swift.' Some corals practise interspecific aggression to open up space: polyps of one species may recognize the tissue of a neighbouring colony and then use specialized long sweeper tentacles to attack or may extrude mesenteric filaments to digest the neighbour's tissues.[22] Or, one colony may inhibit another and maintain a gap between the two without physical contact, presumably by diffusible chemical weaponry. Alcyonacean soft corals on tropical reefs likewise compete with scleractinians, and various seaweeds produce chemicals that they use to inhibit corals when living in close quarters with them. Photographs or maps of coral communities therefore have a mosaic or patchwork appearance that results from long-term aggression and competition (illus. 34).

As sessile, branching colonies of modular polyps that capture and consume plankton borne by the surrounding seawater, filter-feeding corals rely on currents to convey this food, as well as oxygen and nutrients, to them. The water flow increases with the distance away from the slow-moving boundary layer at the seafloor, so it behoves a fixed colony to grow away from the substratum and into the flow's nutritionally richer mainstream (so long as the current is not so strong as to deform or damage the colony).

A special case illustrates this point. Like other deep-dwelling filter feeders, gorgonians in the abyss grow slowly, a consequence both of their nutritionally dilute environment and the often flat, featureless plain where hard substratum is scarce and food-bearing currents are slow. But a colony of *Chrysogorgia* sp. that was photographed repeatedly by the Russian *Mir* manned submersibles over ten years grew some four times faster than corals attached to deep-sea cables lying flat on the bottom.[23] This colony, however, was attached not to those cables but to the uppermost rail on the deck at the bow of RMS *Titanic* (the towering, windy spot famously frequented by the characters

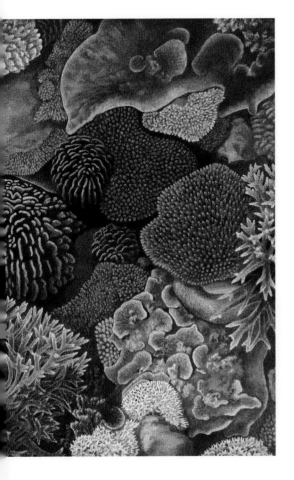 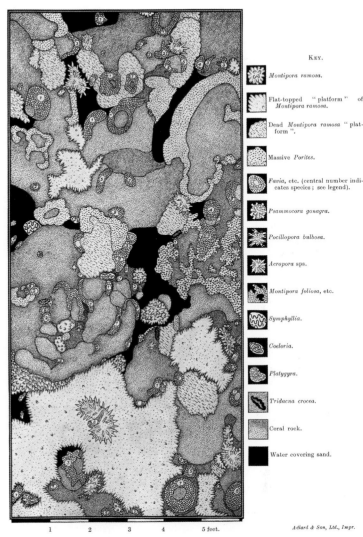

34 and 35 The patchwork appearance of a reef coral community on the Great Barrier Reef is evident in T. A. Stephenson's oil painting published in 1946 (left), and in a large-scale map drawn in ink for his 1935 report with K. Cole on the 1928–9 Great Barrier Reef Expedition.

portrayed by Leonardo DiCaprio and Kate Winslet in the 1997 film) (illus. 36). Even inside the ship lying on the bottom about 3,800 m (12,500 ft) deep, unidentified gorgonians or bamboo corals are growing on a chandelier above the grand staircase, a testament to the opportunism of these beings that require hard substrata and life-giving currents.

Additional environmental factors impact the growth and form of corals: for species harbouring photosynthetic endosymbionts, the most abundant prey is photons. Down deep, where sunlight has been dimmed during its long passage through the optical filter of seawater, scleractinian colonies flatten horizontally or assume an open framework to present more surface area and avoid self-shading to capture what remains of

Opposite page: 36 A fast-growing colony of the gorgonian Chrysogorgia *on the bow rail of RMS* Titanic, *photographed by the Russian* MIR-I *submersible. Above: 37 Branching staghorn corals show heliotropic growth as they grow outward, away from the shade under a plating laminar coral, and then upward, toward the sun. The plating coral extends horizontally to fill the spaces between the staghorns' branches in this competitive interaction.*

downwelling light for photosynthesis by the symbiotic algae. Like understorey plants on land, submarine corals may grow heliotropically, out of the shade and towards the sun, their growth trajectory recorded in their stony skeleton (illus. 37).

The modular nature and growth of coral colonies is reminiscent of a musical fugue, with its copies of a subject and its recursive, interwoven structure. The rate and direction of growth may change when environmental conditions do, leading to complex morphologies. Such plasticity is afforded by indeterminate colony growth through the addition of modules (that is, polyps, with variable spacing between their corallites) that shape the colony according to its environment, as demonstrated in field transplant studies. Decreased water flow affects the availability not only of planktonic prey but also the deposition of smothering sediment in turbid back-reef and shady mangrove areas, necessitating an open architecture to avoid both clogging and self-shading. Conversely, a more robust and compact morphology is adopted to resist waves breaking on the reef flat, a morphology that becomes more

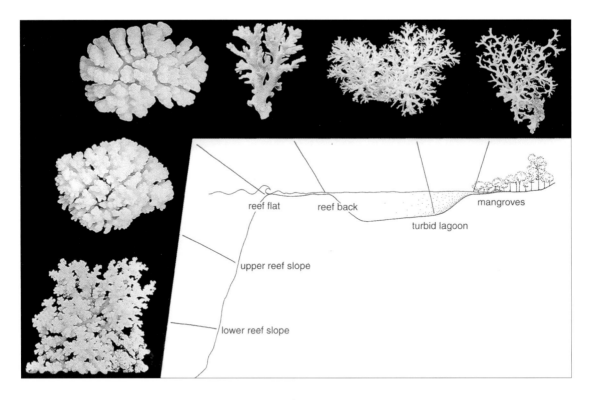

open in deeper water on the reef slope, where light is lower and there is less wave action (illus. 38).

Consequently, for a eurytopic species such as *Pocillopora damicornis* that occupies diverse habitats as 'ecomorphs' having different locally adapted forms, the already ambiguous coral becomes a shape-shifter, changing its morphology spatially and temporally and confounding classification and assessment of biodiversity. Such plasticity confused early taxonomists, one of whom wrote about *Pocillopora*: 'we have here a chaos of forms.' But it is precisely this phantasmagoric flexibility that came to be seen as allowing scleractinians to be so adaptable and to survive in so many different circumstances. No wonder, then, that colony form is but one of the characters used by contemporary taxonomists, who are turning increasingly to molecular genetic analyses.

In *P. damicornis*, such analyses have shown the array of described ecomorphs to correspond

38 Different colony morphologies in Pocillopora damicornis *are suited to habitats varying in solar irradiation, wave action and suspended sediments.*

to several genetically distinct lineages. Thus, *P. damicornis* actually may be several 'cryptic species' each having its own range of eco-phenotypic variability,[24] yielding disparate colony forms that do best under different circumstances. Molecular studies also show the incidence of both cryptic species and interspecific hybrids (and even chimeras) in other scleractinians, so that delineating species to assess evolution, biogeography and biodiversity in these corals is fraught with difficulty.[25]

Early scleractinians may have benefited from symbiosis with dinoflagellates, those golden-brown unicellular algae called zooxanthellae, whose ancestors appeared at about the same time in the fossil record.[26] This association – with microscopic algae less than 10 μm (micrometres) in diameter living as endosymbionts inside the host's gastrodermal cells – unifies carnivorous polyp with photosynthetic alga. The zooxanthellae capture the energy in sunlight to fix inorganic carbon dioxide from seawater into organic compounds to be used in the host animal's metabolism. This cooperative combination in one holobiont provides the flexible nutritional possibilities that, together with the enhancement by the algae of the host's calcification, in large part account for the success of reef-building corals (illus. 39 and 40).

In some corals that brood their larvae internally, the algal symbionts are inherited maternally, being passed from female parent to egg. Larvae that develop from eggs released by broadcast spawners usually must acquire the algae from the environment. The density of the algae living inside the coral host can be extremely high, up to several per gastrodermal cell but collectively a million or more algal cells per square centimetre of colony surface.

In corals these zooxanthellae predominantly are dinoflagellates in the genus *Symbiodinium*[27] (illus. 41), which have proven to be exceptionally diverse genetically and are the subject of studies on the physiological and ecological relevance of their genetic variation. For example, some clades (genetic lineages) of *Symbiodinium* appear more tolerant of high temperature and bright sunlight than others, perhaps conferring resistance to the thermal stress-related loss of algal symbionts – 'coral bleaching' – that has devastated some reefs and thereby offering a glimmer of hope for corals in an ever-warmer era of global climate change, a topic for the final chapter.

Also, stimulated partly by incidences of diverse diseases of corals, the finding of antimicrobial activity in coral tissues, and growing knowledge of the coral immune system that has elements in common with that in humans, studies of the nature of the associations

39 and 40 (left) Natural-colour photograph in white light of the crown of an excised Stylophora pistillata *polyp showing golden-brown zooxanthellae inside the polyp and emerging from cut tissues, and cnidocytes ('stinging cells') in the tips of the tentacles; (right) induced fluorescence photograph of the same polyp showing red-fluorescing chlorophyll in the zooxanthellae and Green Fluorescent Protein in the host tissue. The polyp is the size of a pinhead (~1–2 mm in diameter) and the zooxanthellae are microscopic (8–10 μm in diameter).*

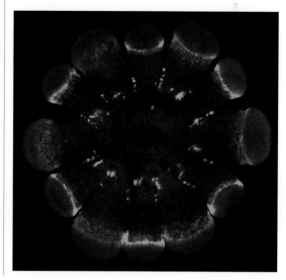

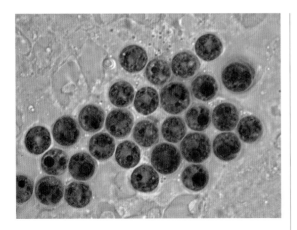

41 Photomicrograph of Symbiodinium *cells isolated from a cnidarian host and growing in culture.*

of bacteria and viruses with corals are proliferating in the new era of metagenomic studies of species associations in the web of life, with an emerging emphasis on this 'microbiome'.[28] One example is the association of certain reef corals with bacteria that are able to fix atmospheric nitrogen gas into organic compounds.[29] If these compounds are available to the host or its zooxanthellae, they provide an important source of organic nitrogen, such as amino acids used to build proteins, important because most of the organic carbon available from zooxanthellae is lipid- and carbohydrate-rich 'junk food', rich in calories to burn as fuel but poor in nitrogen with which to grow.

A more recent finding is the stable predominance of spirochaete bacteria among the microbial associates of healthy colonies of *Corallium rubrum*, a unique situation in the Anthozoa.[30] Although certain spirochaetes cause Lyme disease and syphilis in humans, other members of this small group form crucial nutritional symbioses in the digestive tracts of termites and ruminants. The role of spirochaetes in the well-being and metabolism of red corals remains to be established.

2 On the Nature of Corals

> The flower of that supposed plant was nothing, in truth, but an insect resembling a small sea anemone or octopus [*poulpe*].
> Jean-André Peyssonnel, *Traité du corail* (1726)

Aristotle did not mention red coral or *kouralion* but did consider the related soft coral *Alcyonium palmatum*, the *main de mer* (literally 'hand of the sea', or 'dead man's hand' or *main de larron*, the 'robber's hand') (illus. 44). He called it *pneumōn* (lung) for its spongy internal structure[1] and placed it among the 'dualizers' that were neither quite one thing nor another. The *main de mer* eventually would have a hand in the recognition of the true nature of corals. Aristotle expressed the broader difficulty (in *The Parts of Animals* of *c.* 350 BCE, and in Book VII of *History of Animals*) of classifying marine organisms that fall outside his established categories of animals, and between plants and animals in their characteristics.[2]

This 'middle nature' of coral had Theophrastus (*c.* 371–*c.* 287 BCE) placing it both in his *De lapidibus* (On Stones) and *Historia plantarum* (Inquiry into Plants). Whereas in the former work it was a stone that grew like a root in the sea, in the latter it was unequivocally a well-known sea-plant growing near the Pillars of Hercules (Strait of Gibraltar) that turned to stone.

Coral Conundrum

In addition to the precious Mediterranean *Corallium rubrum*, Theophrastus, looking further afield, mentioned a 'petrified Indian reed', perhaps the octocoral now called *Tubipora musica* after its resemblance to pan pipes or organ pipes (illus. 42 and 43). Both *Corallium* and *Tubipora* have calcareous skeletons biochemically tinted red, but inside-out relative to each other: in precious coral the skeleton is the inner axis of the colony covered by soft tissues and the polyps, while in the 'organ-pipe' coral it is an array of tubes, each surrounding a soft polyp. Both species later would also feature in the recognition of the animal nature of corals.

Also, according to Theophrastus, in the 'Gulf of Heroes' (Gulf of Suez), there were tree-like plants 'like stones so far as they project above the sea', but green and with conspicuous 'flowers' when underwater, being some 3 cubits (1.35 m/4 ft 5 in.) tall. In short, he also knew of the scleractinian coral reefs of the Red Sea. In his *Geography*, written in the first decades of the first century CE, Strabo too told of underwater 'trees' growing along the entire coast of the Red Sea, to him all the more unusual because the land above the water had no such vegetation. According to Pliny, such submerged trees ripped the rudders from ships in the Indian Ocean.

Dioscorides, that first-century CE Greek surgeon in Nero's army, likewise considered red coral a marine plant (noting that some called it Lithodendron, or 'stone tree'). In his *De materia medica* he described its uses, which included its cooling, anti-inflammatory effect and, not surprisingly given its colour, giving it to patients who were spitting up blood. Dioscorides knew also of the more distantly related antipatharians (black corals), which had medicinal properties similar to precious coral's. Other therapeutic uses of coral are described in subsequent European texts,[3] and in Chinese, Tibetan, medieval Arabic, and Japanese, volumes of *materia medica* (illus. 45).[4]

Pliny the Elder in his *Natural History* solidified the botanical character of corals, including both scleractinians of the Indian Ocean

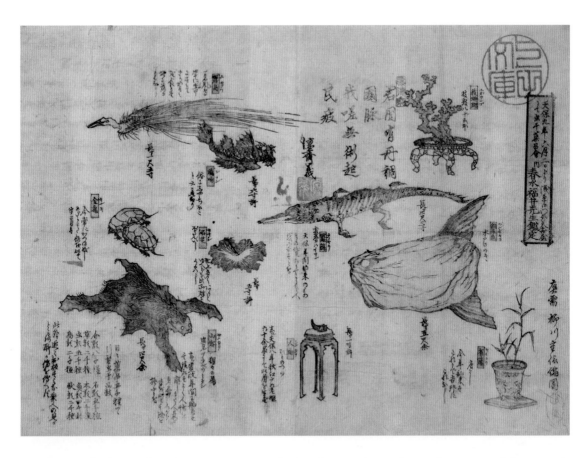

Opposite page (top): 42 and 43 The organ-pipe octocoral, Tubipora musica: (left) Augusta Foote Arnold, The Sea Beach at Ebb Tide, *1903, from Ernst Haeckel,* Arabische Korallen *(1875); and (right) William Saville-Kent, figs 6–8, detail of chromo plate X in* The Great Barrier Reef of Australia *(1893); (below): 44 Marguerite Marie Martin's painting of white and purple* Alcyonium palmatum *colonies, seen in* The Flowery Rock and Pink Wrasses *(1936) in the aquarium of the Musée Océanographique de Monaco, illustrates why this soft coral is called* main de mer *or 'dead man's hand'. Aquarelle and gouache. Above: 45 Shigenobu Yanagawa,* Yakuhinkai syuppinbutsu (Pharmaceutical Exhibit Products), *1840. In addition to what is probably a branch of Mediterranean red coral in the top right, the items include a dried ocean sunfish (*Mola mola*) below that, then a rhinoceros horn, and at lower left an orangutan pelt.*

– those undersea forests that menaced ships – and Mediterranean coral, noting its green, soft, shrublike nature underwater and its white berries, but also that the plant hardened and turned red when exposed to air. The Mediterranean coral that Pliny was describing is *Corallium rubrum*, which is related to the diverse group called gorgonians. Its outer soft, living coenenchyme (formerly called the *écorce*, or bark) toughens when dry. Removing this layer reveals the central red axis that is polished and used in making jewellery.

The foregoing version from Pliny is the one often seen. Actually, he wrote that it was the white berries that hardened and turned red, resembling in size and colour the berries of cultivated dogwood. Pliny also mentioned antipatharian black coral from the Red Sea, calling it *iace*. Leo Wiener (1862–1939, father of the cybernetician

Norbert Wiener) told us that *iace* is from an Arabic word not only for this coral but also for 'black pearl, a plant, the black kernels of which are made into beads'.[5] Might Pliny have conflated such black kernels with carved beads of red coral that resembled dogwood berries?

In his *Natural History of Norway* (1755), Erik Pontoppidan, Bishop of Bergen, also noted the confusion of coral beads and supposed coral berries: 'the little beads, made of coral (they not being as some imagine, fruits or little berries growing thereon)'. Pontoppidan went on to suggest that there might likewise be a fashion market for white beads made from the cold-water coral abundant in Norwegian seas, almost certainly the scleractinian *Lophelia pertusa*. Pliny also discussed the trading of Indian pearls for red coral, noting that each fetched a similar price and that the value of things depends on personal tastes.[6] Certainly red coral was clearly distinguished from reef-building scleractinians by the twelfth century, when the author of a Persian treatise wrote that 'the infidels (*kāferān*) of Cathay prefer coral to jewels' but that the white (scleractinian) coral near the port of Hormuz was 'good for nothing'.[7]

46 *Khadiravani Tara at Work in Tibetan History, Tibet, 19th century, ground mineral pigments on wood. This detail shows red coral being collected on mountainous islands with orb-shaped gems in an ocean inhabited by dragons.*

Mediterranean red coral made its way to China before the Common Era. The word for coral, *shanhu*, appeared in a Han dynasty poem written during the mid-second century BCE,[8] when coral was regarded as a mineral. As knowledge of coral expanded, it came to be seen as a tree, as in the eighth-century CE poem 'Yong Shanhu' by Wei Yongwu:

> Red tree with neither flowers nor leaves,
> Neither a stone nor a jewel,
> Where did people find it?
> They say it was born on a stone on Pengla Mountain over the East Sea.[9]

Thus coral came not from the natural world of humans but from the mythical realm of gods and immortals, Mount Penglai (illus. 46).

Writing at about the same time, Zhong Ziling placed the coral tree under the sea, and gave it mineral properties and an air of mystery, too, in 'Shanhu Shu-fu':

> Coral is born in the great ocean where fish and dragons live. Receiving the spirits of heaven and earth, absorbing nutritious liquid on the sea floor, it becomes the greatest jewel of all. It grows in black mud on the sea bed, yet it is never tinted black, just as the jewel does not lose its purity. It has the form of a tree, with branches, although it has no leaves.
>
> Soon it sprouts, firmly develops roots and starts to grow. It hardens gradually, like ice developing, and its red colour resembles a will-o'-the-wisp quietly burning. Branches grow like red gemstones . . . and its bright colour stands out like the sheer red crest of a rooster.[10]

In 1857 the marine biologist Henri Milne-Edwards (once Cuvier's student) gave a history of the nature and discovery of red coral in the West in the first volume of his *Histoire naturelle des coralliaires, ou polypes proprement dits* (Natural History of Corals, or Polyps, Specifically), used widely in subsequent shorter histories. After relating the well-known examples of Theophrastus, Dioscorides and Pliny the Elder, Milne-Edwards jumped to the sixteenth century. He opined that Renaissance naturalists did not advance the knowledge of red coral. Still, the collection of coral for jewellery and other ornamentation in curiosity cabinets did increase then, under growing demand from wealthy princes.

Several sixteenth-century European botanists described and figured diverse corals not mentioned by the ancients and which turned up later in the *Systema naturae* of Linnaeus, who also initially classified them as stony plants. The ancient transformation of a soft submarine shrub to a hardened branch in air continued to perplex naturalists in the sixteenth and seventeenth centuries. Naturalists sailing with Mediterranean divers to observe living corals (and not just their dried remains in curiosity cabinets) concluded that coral was as hard underwater as it was in air. None of these reports changed the view of corals as submarine plants.

In 1671 the Sicilian botanist Paolo Boccone confirmed that the lower portions of the coral branches were already stiff underwater, whereas the extremities were flexible, and the covering 'crust' was soft and slippery when the coral was fresh, unlike the rough covering in dried specimens. Removing the soft and supple *écorce* revealed the hard, red central axis. He disregarded the descriptions of 'fleurs du corail' by apothecaries in Marseille, despite his own study of the star-shaped pores that contain the polyps in red coral, and, finding no 'flowers, leaves, or root' in his specimens, concluded that coral was not a plant but some sort of stony concretion, that is, a mineral.[11]

Paracelsus, the early sixteenth-century Renaissance physician and alchemist noted for his use of medicinal chemicals and minerals, in his *Book of Vexations* also had viewed coral as a stone that coagulated in the sea (prompted, perhaps, by his medical interests in the formation of kidney and urinary bladder stones). Paracelsus also gave a recipe for counterfeit precious coral, which involved grinding cinnabar with egg whites and forming the paste into branches. Georgius Agricola, too, in *De natura fossilium* of 1546, the first textbook of mineralogy, presented precious coral as a hardened fluid, a 'concretionary juice'.

Michael Maier, an alchemist, Rosicrucian and physician to Holy Roman Emperor Rudolf II, included Sicilian red coral in his 1617 sourcebook of alchemical emblems, *Atalanta fugiens, hoc est, emblemata nova de secretis naturae chymica* (Atalanta Fleeing, that is, New Chemical Emblems Regarding the Secrets of Nature), written in humanistic Latin. (illus. 47) Maier, drawing from ancient European and Arabic sources, likened the philosopher's stone to curative stones (including pearls and amber), and especially to plantlike coral. He drew his analogy from their both growing in water but taking their nourishment from the earth, and their eventual hardening when exposed to air; from their redness ('coraltincture'), the colour of blood and health; and from their medicinal properties:

> Just as coral may be used for several potent medicines, so the Philosophical Coral bears the powers of all herbs, because it possesses as much curative power as all herbs together . . . It is the Philosophical Coral, vegetable, animal and mineral, which keeps itself hidden in the vast sea and is not recognized.[12]

In his *Coralliorum historia* (1669) the physician Johann Ludwig Gans, although also quoting the elusive Orpheus on the association of corals with marine plants, perpetuated the alchemical view of corals as spontaneously hardening gelatinous materials that steer their own shrubby shapes. Such self-organizing structures that are not biogenic *do* exist in nature – crystals. Luigi Ferdinando Marsigli sailed with coral fishers on the coast of France in the first years of the eighteenth century and collected corals from depths as great as 150 fathoms (275 m/900 ft). He viewed corals as marine concretions that often grew downward from the roofs of submarine caverns, and sought to banish them from the vegetable kingdom: 'they did not spring from seed but were formed in the same way as crystals found in underground caves.'[13] (illus. 48) A century and a half later, Flaubert wrote of 'fleurs de fer', iron flowers or *flos ferri* – crystals of aragonite associated with iron ores – when describing the confusing nature of corals in *The Temptation of Saint Anthony*. The complexity of the seemingly colonial architectures of some such crystals belied their inorganic mineral nature (illus. 49).

François Pyrard de Laval, describing his shipwreck in the Seychelles in 1602, would tell of meeting

> with many branches of a certaine thing which I know not whether to terme Tree or Rocke, it is not much unlike white Corall, which is also branched and piercing, but altogether polished; on the contrary, this is rugged, all hollow and pierced with little holes and passages, yet abides hard and ponderous as a stone.[14]

Here the writer is describing reef-building corals and likening them to the polished white variant of *Corallium rubrum* with its mineral composition but botanical form.

47 Merian the Elder's copperplate engraving for Emblem XXXII in Michael Maier's Atalanta Fugiens *(1617). The Latin motto that accompanied the icon reads: 'As coral grows under water, and, exposed to air, gets hard, so also the stone.'*

Representative of late seventeenth-century botanical scholars, Joseph Pitton de Tournefort included stony corals and gorgonians among anomalous marine plants in his *Éléments de botanique* (1694). In his memoir of 1700,[15] he clearly figured a gorgonian ('*Lithophytum*') (see illus. 2) with its seemingly woody inner axis (actually, the proteinaceous complex is gorgonin), covered with a rind inhabited by the polyps. John Ray mentioned Georg Eberhard Rumphius's calcareous lithodendrons and black corals among the plants of the Indonesian island of Ambon in the appendix to the second volume (1693) of his *Historia plantarum*, an important early effort towards modern taxonomy. Ray classified corals, together with the marine macroalgae (seaweeds), as flowerless plants. This was the situation in the seventeenth century, with no hint yet of the taxonomic upheaval to come.

Earlier, in 1599, Ferrante Imperato, a Neapolitan apothecary and naturalist, had published *Dell'historia naturale*, illustrated from his own botanic garden and chamber of curiosities. According to Milne-Edwards,

a AMSTERDAM aux Depens de la COMPAGNIE 1725.

Opposite page: 48 The frontispiece, engraved by Matthys Pool, and exceptionally in colour, from Sir Joseph Banks's copy of Marsigli's Histoire physique de la mer *(1725). Neptune is shown on a seabed littered with marine life, including branches of red corals, which grow in caves or under ledges, like the crystals Marsigli at first believed them to be. Above: 49 The biomorphic, seemingly colonial architectures of flos ferri crystals belie their inorganic mineral nature. Aragonite (Var: Flos Ferri): $CaCO_3$, from Styria, Austria.*

Imperato considered most corals in his collection to be marine plants but thought that the tubes in *Tubulara purpurea* (now the familiar *Tubipora musica*, the alcyonacean organ-pipe coral) might have been built by a marine animal in a manner akin to the cells built in hives by bees. But Imperato didn't have it quite right: unlike mobile bees, which live independently of the cells in their hives, sessile polyps never leave the colony. Nevertheless, John Ellis, who eventually would convince Linnaeus himself of the animal nature of corals, credited 'Imperatus' with this early recognition in *The Natural History of Many Curious and Uncommon Zoophytes* (1786).

The dominant opinion by the start of the eighteenth century remained that corals were plants, largely because of the arborescent form of their colonies, but attention would soon shift to the polyps. Even Marsigli had come to see corals as plants, observing the cycle of expansion and contraction of what he called *fleurs du corail* (white coral 'flowers' being botanically consistent with Pliny's description of corals as having white 'berries'). Within six months of his call to banish coral concretions in caves from the vegetable kingdom, he recanted and placed red coral and various madrepores among the 'Plants Pierreuses' (stony plants) in his *Histoire physique de la mer* (Natural History of the Sea) of 1725. Elsewhere in that volume he retained the category 'Lithophytes' for gorgonians to be consistent with older usage, but with reservations: 'The ancients have given them the name of Lithophyton, which in my opinion scarcely suits them, having no natural connection with stones.' Instead he viewed these as 'Plantes presque de bois', akin to 'woody plants' (illus. 50).

But the affable and generous John Ellis would credit not only Imperato but also Georg Eberhard Rumphius for providing earlier evidence that corals actually were animals. Rumphius did not include corals in his *D'Amboinsche Rariteitkamer* (Ambonese Curiosity Cabinet, 1705), putting them instead in his *Amboinsche Kruidboek* (Ambonese Herbal), published between 1741 and 1755, where he considered them as plants 'in the Sea, and which have a mixed nature of wood and stone, and which are called Sea-trees or Coral-plants'.[16] Nevertheless, it was here, apparently, that Rumphius (according to Ellis, who did not cite the *Herbal* explicitly) clearly described the solitary coral *Fungus saxeus* or

Madrepora fungites (now *Fungia fungites*) as having a thick, viscous covering when alive in the sea. The coral also had

> innumerable oblong vesicles [tentacles] ... which appear alive under water, and may be observed to move like an insect: that as soon as the Coral was taken out of the sea, and exposed to the air, all the mucous part, with the little vesicles, shrunk in between the erect little plates, or lamellae, and disappeared; and in a short time, like the Medusae, or Sea Jellies, melted away, leaving behind them a most disagreeable fetid smell ... All the other Zoophytes there, when they are fresh, are possessed by a gelatinous animal of a fishy nature.[17] (illus. 51 and 52)

After commenting at length on the beauty of living corals that were by then known to be animals, Joseph Beete Jukes, aboard HMS *Fly*, reiterated Rumphius's comments regarding rotting

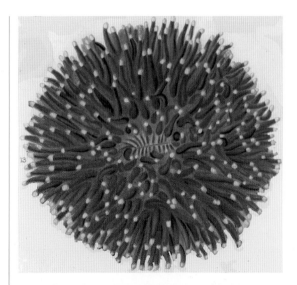

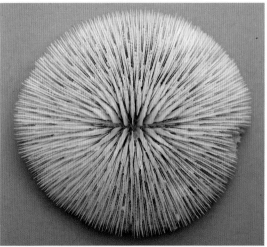

Opposite page: 50 Hand-coloured variant of Matthys Pool's engraving Tab. [tabulum] XXII from Sir Joseph Banks's copy of Marsigli's Histoire physique de la mer *(1725). By this time Marsigli considered red coral to be a marine plant (not a crystal growing in a cave). The middle engraving paired a terrestrial green plant growing upward from a rock, with a red coral beneath it, growing downward toward the centre of the earth, both to link and distinguish the plant and the coral. At bottom is the salabre (a device used to collect corals), mislabelled 'Engin' in the engraving. Right: 51 and 52 A living specimen of a mushroom coral,* Fungia, *with its motile tentacles (top). W. Saville-Kent, fig. 13, detail of chromo plate VI, in* The Great Barrier Reef of Australia *(1893). Cleaned skeleton of* Fungia, *showing the calcareous septa and costae (Ellis's 'lamellae') that would be covered with tissue in a living specimen as above (bottom).*

corals: 'The smell of the animal matter, also, while the corals are drying, is most sickening.'[18]

The sixteenth-century chimeric construct 'zoophyte' neatly encompassed the ambiguity of corals, for 'they are of a dual nature, neither completely animal nor plant', according to Edward Wotton.[19] So, too, had 'lithodendron' and 'lithophyte' earlier, before the intrusion of zoology into the classification of what had been seen as stationary stones and plants.

It was Jean-André Peyssonnel who fully recognized the importance of the polyp in determining the nature of corals. He saw the similarity between the supposed white flowers of red coral and other soft octocorals and even medusae, and replaced those *fiori bianchi* or *fleurs du corail* of Marsigli with sentient 'insects' that not only contracted on exposure to air but expanded or contracted when touched or experimentally subjected to acidified or warmed seawater in a vase (illus. 53). Warming also generated the 'very disagreeable' odour of putrefying animal tissue, as Ellis noted that Rumphius had found. But it was the coral insects' quick responsiveness to stimulation – their irritability – that prompted 'a metamorphosis as surprising as any in Ovid': the transformation of living beings from one kingdom of life into another. Those few naturalists who did observe a living coral colony would have seen it in air or a bucket soon after its collection, when the polyps first contracted to small white blobs resembling berries. Peyssonnel himself saw them when, like Boccone and Marsigli before him, he went to sea with coral fishers in 1723.

Although Peyssonnel is now widely credited with proving that corals are animals, in the day his conclusion was opposed by the influential polymath René-Antoine Ferchault de Réaumur, who argued that coral was neither animal nor plant but rather a stony *product* of a plant. Peyssonnel had corresponded with Bernard de Jussieu and Réaumur, who were uniformly discouraging and ironic in their replies. After reading Peyssonnel's ideas to the Académie des Sciences in 1726 without naming their originator, Réaumur suppressed publication of Peyssonnel's communication to the Académie and argued against its conclusions in his own 1727 memoir.

Subsequently the experiments on freshwater hydras by Abraham Trembley, communicated to Réaumur and later published in 1744, convinced Réaumur of the true nature of these animalcules. With Jussieu (who, also stimulated by Trembley, studied the 'dead man's fingers', the alcyonacean

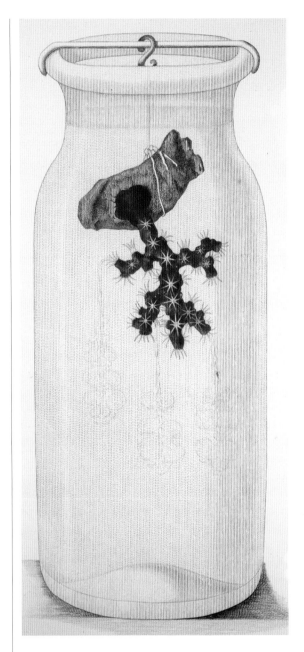

53 *Henri Lacaze-Duthiers's* Histoire naturelle du corail *(1864), detail of plate XII. Peyssonnel's experiments with living red corals in laboratory vessels had convinced him that Marsigli's white fleurs du corail actually were sentient animals similar to a sea anemone or an octopus.*

soft coral *Alcyonium digitatum*), Réaumur named them 'polypes' by analogy with the arms of the octopus (*poulpe*) as recognized explicitly by Peyssonnel. Réaumur also came to apply the term to sea anemones and various corals, with overdue acknowledgement of Peyssonnel. Réaumur, an entomologist, further analogized the architecture of the cup-like support (corallite) of the individual polyps to the cells of bees and wasps, calling the former 'polypiers', or polyparies.

Réaumur's confessional recanting of the nature of corals, lithophytes, *Alcyonium* and so on was in the preface to volume VI of his *Mémoires pour servir à l'histoire des insectes* of 1742. There he professed his respect for Peyssonnel and said that omitting Peyssonnel's name from his own earlier treatise was to protect the author of such a rash idea. In a retrospective analysis to set the record straight, published in the *Annales des sciences naturelles* in 1838, the neurophysiologist Marie-Jean-Pierre Flourens quoted extensively from Peyssonnel's original manuscript and deemed Réaumur's action 'considerate'. Nearer to the time of the original controversy, the Comte de Buffon was less forgiving and wrote in his *Histoire naturelle* (1749) that, 'having long doubted the truth of Mr Peyssonnel's observation, some naturalists, too protective of their own opinions, at first rejected it with a kind of disdain; however, recently they were obliged to reconsider Mr Peyssonnel's discovery.'[20] Buffon, validating Peyssonnel, was unequivocal in his taxonomic placement of corals: 'this way marine plants that first we had ranked among the minerals, then passed into the class of Plant, finally are to remain always in that of animals.'[21]

Soon after Buffon's conclusion, the botanist and marine biologist Vitaliano Donati's opinion that corals were animals was translated and entered into the *Philosophical Transactions of the Royal Society of London* in 1750. However, it changed few minds there; it may not have helped that his opening sentence was 'Coral is known to be a marine vegetation, which in shape nearly resembles a shrub stripped of its leaves.'

Peyssonnel's 'curious treatise', by now some four hundred pages long, sent from Guadeloupe where he had relocated on a royal appointment and continued to study a wider array of corals, was translated into English and read in 1752 to the Royal Society. It appeared in abridged form the next year in the Society's *Philosophical Transactions*, offering a wider forum for his ideas and experimental results. Subsequently, T. H. Huxley's essay 'On Coral and Coral Reefs' (1873) relating this publication and sarcastically describing the French academic establishment's treatment of

54 Fessard, after J.-C. Allais, late 18th-century engraving commemorating Dr Jean-André Peyssonnel, shown holding a jar containing a living colony of red coral.

the original communication helped to cement Peyssonnel's position in history (illus. 54).

Donati refined his views by 1757, leaving no doubt in a letter to Abraham Trembley that a coral is an animal having many heads (the polyps) and that its bone is shaped like a shrub and covered by animal flesh. For our contemporary writer James Hamilton-Paterson, 'An animal in the shape of a shrub is – if nothing else – the exact opposite of topiary, and suitable for a zoological garden.'

Based on proliferating contemporary accounts and opinions, and especially because Ellis sent him illustrations and descriptions intended to show the animal nature of corals, Linnaeus eventually enthroned corals and gorgonians in the animal kingdom, as a mixed bag of 'imperfect' beings in the Lithophyta and Zoophyta, in the tenth edition of his *Systema naturae* (1758). But Linnaeus's stubborn ambiguity was evident in his personal response to Ellis:

> Zoophyta are constructed very differently, living by a mere vegetable life, and are increased every year under their bark-like trees, as appears in the annual rings in a section of a trunk of Gorgonia. They are, therefore, vegetables, with flowers like small animals, which you have most beautifully delineated . . . Yet as they are endowed with sensation and voluntary motion, they must be called, as they are, animals; for animals differ from plants merely in having a nervous sentient system, with voluntary motion, neither are there any other limits between the two.[22]

Even though the animal nature of corals and other polyps was established, the shrub-like form of coral colonies remained a philosophical puzzlement. Voltaire hinted of this confusion in his spoof of the attitude of the French Academicians in his *Dictionnaire philosophique*, first published in 1764, under the entry 'Polypus':

Leuenhoeck raises them to the rank of animals. We know not if they have gained much by it . . . It appeared to us that the production called polypus resembled an animal much less than a carrot or asparagus. In vain we have opposed to our eyes all the reasonings which we formerly read; the evidence of our eyes has overthrown them. It is a pity to lose an illusion. We know how pleasant it would be to have an animal which could reproduce itself by offshoots, and which, having all the appearances of a plant, could join the animal to the vegetable kingdom.

John Ellis, in *The Natural History of Many Curious and Uncommon Zoophytes*, was generally supportive of Peyssonnel and used the terms 'animal' and 'zoophyte' interchangeably. But Ellis went astray in criticizing Peyssonnel's view of the formation of coral colonies: 'How absurd, then, is it to suppose that corals compounded of many such animals [polyps], each upon its cell, do vegetate as plants, because they grow up together in ramified forms.' But Flourens, in his appreciation published a century after Peyssonnel's seminal findings, further credited him with recognizing that corals are 'animaux composés' – 'many animals linked by a common body' – and reminded readers that Trembley's discovery of vegetative reproduction in polyps by budding provided the mechanism for colony formation by corals, derided by Ellis.

Samuel Taylor Coleridge, the Romantic philosopher and one of the English Lake Poets at the turn of the nineteenth century, had been aware of coral's colonial nature, where 'a multitude of animals form, as it were, a common animal.' He also knew how they did it: 'The mere existence of a polypus suffices for its endless multiplication. They may be indefinitely propagated by cuttings, so languid is the power of individuation, so boundless that of reproduction.'[23] The branching form of colonial corals clearly impressed Coleridge: early on, in a notebook entry of 1800, he perceived it in a teaspoonful [of laudanum] ramifying as it ran down the inside of a glass as 'lovely coral-shaped shadows'.[24]

Writing retrospectively in 1776, the abbé Jacques-François Dicquemare reflected that to understand corals, one had to distinguish the polyps from their productions, and those in turn from plants:

> If something were capable of causing an error, it would be the polyps that form corals . . . These little animals seem to unite the animal kingdom to the vegetable, because of their form; their admirable productions, seen as vegetative stones, shade between vegetable and mineral. But it is easy to distinguish between the polyp, the polypary, and the marine plant . . . & the more we observe, the less we will have of the equivocal.[25]

Using much of the same vocabulary, Flaubert reprised this tripartite theme in the middle of the nineteenth century, towards the end of *The Temptation of Saint Anthony* (quoted in my Prelude). A century and a half after that, Osha Gray Davidson, in *The Enchanted Braid* (1998), would coin the knowing neologism 'zoophytelite' to interweave the animal, vegetal *and* mineral natures of coral by adding the stony suffix to grow the etymological crystal. And their ambiguous or deceptive appearance led the contemporary artist Gerardo Stecca to refer to the 'specious morphology' of corals.[26]

Ellis's own lifework, *The Natural History of Many Curious and Uncommon Zoophytes*, was not published until 1786, ten years posthumously. In that volume and elsewhere Ellis applied the old English term 'flower animals' especially to sea anemones, the close relatives of corals. His descriptions of species were arranged by Daniel Solander, a naturalist who had sailed with James Cook and Joseph Banks on *Endeavour* and who also died before the book was published and dedicated to Sir Joseph. 'Ellis & Solander' appear as the original describers of many coral species. Linnaeus was the mentor of 'the ungrateful Solander', who despite Linnaeus's matchmaking did not marry his daughter or even send him any specimens from the *Endeavour* voyage.[27] Linnaeus did not erect a higher level of classification for zoophytes and lithophytes. It seems it was Christian Gottfried Ehrenberg who in 1834 first used 'Anthozoa' to comprise soft and hard corals, sea anemones and other 'flower animals'.

The Maya, too, had seen corals as floral. A ceramic bowl of the Early Classic period (450–500 CE) depicts the maize god in a watery underworld that includes shells of thorny oyster (*Spondylus*) and conch (*Strombus*) and a central image of a skeletal head with emanating floral elements interpreted as corals, being grazed by fishes,[28] as divers on reefs see them today (illus. 55).

Solander and the wealthy Banks ('a gentleman of £.6000 per annum estate'[29]) were well prepared to study corals aboard the *Endeavour* on Cook's first Pacific voyage of 1768–71. Ellis, informed by Solander, wrote in 1768 in a letter to Linnaeus that the explorers were expensively equipped for studying corals using 'all kinds of nets, trawls, drags and hooks for coral fishing; they have even a curious contrivance of a telescope, by which, put into the water, you can see the bottom at a great depth'.[30] Despite these preparations, because the naturalists' time was largely taken up by novel 'fish, Plants, Birds, &c &c.' rather than corals, Banks in his journal 'lamented that we had not time to make proper observations upon this curious tribe of animals'.[31] Neither Cook nor Banks in their journals commented on the biology of corals, which were collected mostly adventitiously, but instead presented coral reefs as navigational hazards. Ironically, *Endeavour* would run aground on the greatest of coral creations, the

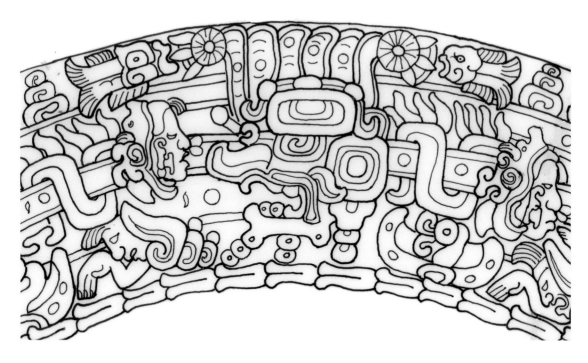

Great Barrier Reef, rising abruptly out of what had been safe, deep water and almost ending Cook's first Pacific voyage.

Once the biological status of corals had been settled, attention focused on their geological manifestation, bridging as reefs do the living and inanimate worlds. Jean-René-Constant Quoy and fellow surgeon-naturalist Joseph Paul Gaimard, among the first to study coral reefs scientifically, were explicit about this conjunction in the title of their 1824 memoir 'on the growth of lithophyte polyps considered geologically'.

The earliest experiences with this geology were as a navigational hazard and craggy deathtrap, perhaps most famously by Cook on his first Pacific voyage (illus. 57). Banks wrote in his *Endeavour* journal that

> A Reef such a one as I now speak of is a thing scarcely known in Europe or indeed any where but in these seas: a wall of Coral rock rising all most perpendicularly out of the unfathomable ocean, always overflown at

55 Detail of a drawing of the lid of an Early Classic Mayan ceramic bowl, 450–500 CE, from Quintana Roo, near the border of Yucatán and Belize. The floral elements on the skeletal head of the water-lily monster are newly interpreted as corals being nibbled by fishes.

> high water commonly 7 or 8 feet [less than the draught of the ship and, depending on lighting conditions, invisible from the deck and thus a nightmare for mariners] and generally bare at low-water; the large waves of the vast ocean meeting with so sudden a resistance make here a most terrible surf Breaking mountains high, especially when as in our case the general trade wind blows directly upon it.

Banks and Cook often read each other's journal, and Cook described the reef in identical terms.

That such shoaling, stony reefs were the constructions of so-called insects, worms or animalcules was by Cook's time well known (and mentioned by Cook himself in 1774, on

56 A high peak of Bora Bora and its surrounding barrier reef.

his second Pacific voyage, aboard HMS *Resolution*: 'Coral rockes were first formed in the Sea by animals'). The prevailing view, later expressed by the ambitious Matthew Flinders based on his experience during his 1802–3 circumnavigation of 'Terra Australis' and extended charting of the 'Great Barrier Reefs', was that the corals' labours began at great depth and proceeded upward, where 'Future races of these animalcules erect their habitations on the rising bank, and die in their turn to increase, but principally to elevate, this monument of their wonderful labours.'[32] Coral reefs and islands, 'where the geologist to-day may study the first-born of creation',[33] also would become firmly embedded in Charles Lyell's *Principles of Geology* (illus. 56).

Illuminating Corals

The surgeon-naturalists Quoy and Gaimard (then sailing on the 1817–20 voyage of the *Uranie* and *Physicienne*) had recognized that reef-building corals grew only in relatively shallow water. The two colleagues discarded the idea that coral reefs grew upward from the greatest depths of the sea, perhaps (according to the naturalist Jean Vincent Félix Lamouroux) eventually destined to congest the vast Pacific with navigational hazards:

> What then could have given rise to the idea that madrepores clog the ocean basins, and raise from the bottom of their abyss the low islands, dangerous for sailors? Scarcely an in-depth study, a passing glance on the work of these zoophytes.[34]

Yet according to Cook and other navigators, the sheer walls of tropical reefs *did* seem to arise out of unfathomable ocean depths. If Quoy and Gaimard were correct regarding the bathymetric distribution of living corals, Cook's observation would require a *shallow* substratum for even oceanic coral reefs to begin to grow on; in the case of those idyllic, annular atolls far removed from any continent, the round rims of submarine volcanoes rising from great depths offered one seeming solution.[35]

Darwin, however, with an insight that would establish his reputation, intuited the shape of atolls differently, as he explained in *The Structure and Distribution of Coral Reefs* in 1842: fringing coral reefs developed around the shallow shores of volcanic mountains that had earlier emerged from the sea, and as the mountains later subsided over geological time, the fringing reefs, growing outward, 'are thus converted into barrier-reefs; and barrier-reefs, when encircling islands, are thus converted into atolls, the instant the last pinnacle of land sinks beneath the surface of the ocean' (illus. 58). To persist, the reefs had to grow upward as fast as the mountains subsided.

As David Dobbs tells the story in *Reef Madness* (2005), Darwin was just back from his *Beagle* voyage when he presented his subsidence theory (six years before publishing it) to Lyell over lunch. Lyell embraced it immediately, dropping his own idea that reefs rimmed rising volcanoes. 'Reefs were not caps atop mountains that had fallen short. They were, as Lyell put it in a letter to [John] Herschel, "the last efforts of drowning continents to lift their heads above water."'[36] Today, the 'Darwin Point' is the drowning

Overleaf: 57 The Endeavour *replica above corals on the Great Barrier Reef.*

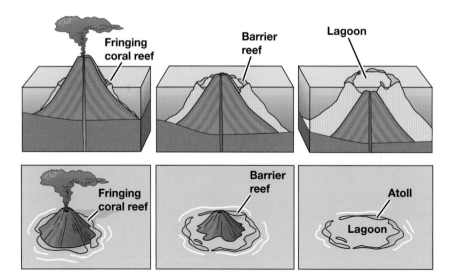

threshold for such reefs and describes the point when their net vertical growth no longer keeps pace with the relative sea level, through geological subsidence, erosion of the reef or rising sea level (which today is accelerating rapidly, as we will see in Chapter Seven).

Darwin's theory did not receive universal plaudits, being rejected most famously by Alexander Agassiz (son of Louis, the great geologist, naturalist and teacher), in part because of Alexander's antipathy towards Darwin, whose ideas about species and evolution eventually supplanted his father's religious perspective. The younger Agassiz argued that coral islands and atolls grew upward from the floor of the deep sea, the corals successively building on the rubble of earlier generations. Were he correct, long cores drilled into coral atolls would contain only coral and reef rock, to the depth of the surrounding seafloor. Darwin's theory, on the other hand, would be supported if such a core contained coral resting on volcanic rock at a shallower depth. In a letter, Darwin even encouraged the wealthy Agassiz to undertake such a drilling project, which he did, but was never able to recover sufficiently long cores.

The opportunity to use newer drilling technology did not come until 1952, in geological

58 Darwin's subsidence theory of atoll formation. According to Darwin (in The Structure and Distribution of Coral Reefs *(1842)), 'both in atolls and barrier-reefs, the foundation on which the coral was primarily attached, has subsided; and ... during this downward movement, the reefs have grown upwards', forming in turn a fringing reef, a barrier reef and an atoll.*

surveys preceding hydrogen bomb testing on Enewetak Atoll in the Marshall Islands. Dobbs reported the result in *Reef Madness*:

> The first cores were clearly reef rock, as expected. As the drill passed the first few hundred feet and out of coral reef depth, the cores changed little. They still appeared to be reef rock ... Finally, at 4,200 feet, the drills hit what was unequivocally basement, a greenish basalt, the volcanic mountain on which the reef had originated ... For more than thirty million years this reef had been growing – an inch every millennium – on a sinking volcano, thickening as the lava beneath it subsided. Darwin was right. Agassiz was wrong.

But why this upward struggle of reef-building corals, and why did they grow only in relatively shallow water? Despite the long comparison of their arborescent form with that of plants, there seems to have been scant attention given to the possible role of sunlight. This oversight came perhaps because some stony corals were known to grow also in the perpetually cold, dark depths of the sea, where extensive mounds of deep-sea corals are being rediscovered today (illus. 59).

Indeed, deep-sea corals grow beneath Arctic and Antarctic seas: according to Jules Michelet's English translator, 'Amidst all the freezing horrors of the Antartic [sic] pole, not far from the volcanic Erebus, Captain James Ross found living coral insects at the depth of a thousand fathoms below the surface of the frozen sea.'[37] (illus. 60) Ross's corals appear to have been mostly gorgonians and hydrocorals (and some bryozoans). But Michelet would write radiantly of tropical reefs as

59 A living cold-water coral (almost certainly Lophelia pertusa) that Pontoppidan included in his Natural History of Norway, Part I (1755), plate XIV, fig. A, and that 'expanded like a flower in full bloom', had a distinctly botanical rather than a cnidarian appearance.

> an animal multitude, whose perpetual exudation of mucus continually raises the circle higher and higher, to low water mark; no higher, or they would be dry; no lower, because they would lack the light. If they have no special organ with which to perceive the light, it circumfuses, penetrates, permeates their whole being. The glowing sun of the tropics, which traverses right through their transparent little frames, seems to have for them all the irresistible attraction of magnetism. When the tide ebbs and leaves them uncovered, they, nevertheless, remain open, and drink in the vivid light.[38] (illus. 61)

Likewise, the formidable Ernst Haeckel, in his little book *Arabische Korallen* (Arabian Corals) of 1876, wrote that despite their lack of eyes, corals can distinguish between light and dark. He was among those who had observed microscopic chlorophyll 'granules' in infusoria, but did not mention them in *Arabische Korallen*.

He deliberately conflated anthozoans and flowering plants in the illuminated initial letter 'A' of that book, so had he known of microscopic golden-brown algal cells in corals, he surely would have written of them there. Apart from the puzzling responsiveness manifested as the polyps' expansion and contraction that had led Philip Henry Gosse to call sea anemones 'blossomed beauties' when he visited them in his aquarium by candlelight, Haeckel offered no physiological role for sunlight in corals (illus. 62).

Understanding the critical role of light in the biology of some cnidarians would have to wait until the last quarter of the nineteenth century. Some invertebrates, including sea anemones, had just been found to have chlorophyll, assumed to be a special form of the pigment of animal

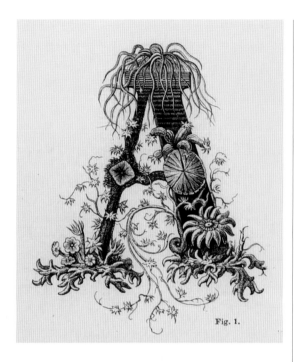

Opposite page: (top) 60 HMS Erebus *and* Terror, *depicted in an 1841 watercolour by J. E. Davis, Beaufort Island and Mount Erebus. Discovered 28 January 1841, with the volcano in the background; (bottom):* 61 *'If they have no special organ with which to perceive the light, it circumfuses, penetrates, permeates their whole being. The glowing sun of the tropics, which traverses right through their transparent little frames, seems to have for them all the irresistible attraction of magnetism.' Jules Michelet,* The Sea *(1864). Above:* 62 *The illuminated first letter of the text in Ernst Haeckel's* Arabischen Korallen *(1876) is an example of his sometimes stylized rather than realistic approach to illustration, here coloured by the old floral view of Anthozoa.*

origin. According to T. H. Huxley as late as 1878, 'Certain animals . . . possess chlorophyll, but there is no evidence to show what part it plays in their economy.'[39] Huxley recognized photosynthesis in plants, which produced organic matter and oxygen (that life-giving gas of Priestley and Lavoisier), and its requirement for both chlorophyll and light, but he did not make the coral connection.

Not long after Huxley wrote, and less than six months before Darwin died in 1882, Karl Brandt and Patrick Geddes independently published, within two months of each other, papers recognizing the algal nature of microscopic 'yellow cells' that occurred in various invertebrates, including sea anemones. The following year, Brandt illustrated the yellow cells that he isolated from diverse invertebrates, including a scleractinian coral, and erected the genus *Zooxanthella* to house them. These 'zooxanthellae' turned out to be a taxonomic mélange, but the name persists as a catch-all for those species of dinoflagellates a few micrometres in diameter that enter into many marine symbioses. Geddes, a social reformer and environmentalist, saw 'reciprocal accommodation' in these associations, likening them to 'animal lichens', albeit in a more complex symbiosis, 'unique in physiology as the highest development . . . of the reciprocity between the animal and vegetable kingdoms'.[40] Brandt's finding the algae in reef corals was scarcely mentioned in subsequent accounts in the late 1800s.

In 1887, C.F.W. Krukenberg described the pigments of several scleractinian corals from the Red Sea, which consistently included the same yellowish-brown colour as the 'yellow cells' in sea anemones,[41] but he seems not to have made the link to algal endosymbionts. As late as 1893, in his book on the Great Barrier Reef that included his experimental observations on corals, William Saville-Kent (who, like Geddes, had been Huxley's student at the Royal School of Mines) did not mention the possible presence of algal endosymbionts in a colony of *Euphyllia glabrescens* and other corals that he studied and illustrated. There,

> certain of the polyps were projected underneath and were shut out from the light

by the surrounding coral growths. In this completely shaded position the tentacles were transparently white . . . Where light only partially fell on them, the tentacles were sage-green . . . while, throughout the area fully exposed to sunlight, all tentacles were dark brown . . . In all the genera named, the polyps screened from light were similarly bleached, after the manner of sea-kale or celery.[42] (illus. 63)

Saville-Kent had written in his earlier *Manual on the Infusoria* (1882) that many of these 'animalcules' contained 'chlorophyll-granules', but he did not make the connection that this pigment underlay the effect of illumination on the colour of what he explicitly called coral-animals, even though he likened the process to bleaching in a seaweed and a garden vegetable! Nor, apparently, did he know of Sydney J. Hickson's book *A Naturalist in North Celebes* (1889), where Hickson suggested that the greenish-brown colour of corals derived from an animal 'substance which is analogous in function to the chlorophyl of plants'.

The first explicit suggestion of the nutritional role of the algal endosymbionts in corals may be in 1898, in a paper where J. Stanley Gardiner attributed coral photosynthesis to unicellular algae inhabiting [the gastrodermal cells of] the digestive cavity.[43] He reported measuring the production of oxygen gas, O_2, as a by-product of photosynthesis in illuminated corals (as Geddes had done for sea anemones) and suggested that photosynthetic 'assimilation of carbon from carbonic acid gas [CO_2] is a very important factor in the growth of corals and hence in the formation of coral reefs.'

With increasing conviction, Gardiner in 1901 claimed that the more vigorous growth of well-lit corals was a consequence of their harbouring 'commensal algae'.[44] By 1903 he was ready to generalize, based on microscopical sections of two exceptional species, that 'They have no *commensal algae*, such as *are found in most reef corals*'[45] (emphasis mine). In short, by the turn of the twentieth century, algal symbiosis had been recognized as the norm on coral reefs, and corals would again become zoophytes (animal–plants), but for a reason unsuspected when they were thus named. For the next century and more, coral–algal symbiosis has pervaded coral research.

Gardiner was instrumental in organizing what would become the British Association for the Advancement of Science's Great Barrier Reef Expedition to the Low Isles off Cairns, under the leadership of C. M. Yonge in 1928–9.[46] The extensive scientific publications of the expedition's results were an unprecedented contribution to the knowledge of environmental effects on the physiology of corals.

Yonge, who had never seen a coral reef and was advised by the experienced Gardiner before the expedition, nonetheless eventually concluded that the endosymbiotic algae played no role in the nutrition of the corals, but

> If we must have a use for the plants [algae], then I think that the speed with which they dispose of the waste products from the corals increases the efficiency of the latter, while they certainly provide abundant supplies of oxygen, without which it is just possible that such immense aggregations of living matter which constitute a coral reef . . . could not originate and flourish.[47]

True enough, but Yonge was summarily discounting Gardiner's view, expressed in

Opposite page: 63 Chromo plate IX from W. Saville-Kent, The Great Barrier Reef of Australia *(1893). The brilliant colours of reef corals arise from the green and yellow-brown photosynthetic pigments in the symbiotic algae and from a wider spectrum of pigments in the tissues of the animal host.*

GREAT BARRIER REEF CORALS.

Gardiner's own extensive report of an expedition to the Maldives and Laccadives, that while corals do eat, 'the vast majority [of reef corals] undoubtedly obtained their nutriment principally through the agency of their commensal algae.'[48]

It was another forty years before the experimental demonstration by Leonard Muscatine and others that not only did the algae 'dispose of the waste products from corals', but moreover they could photosynthetically fix them into organic molecules and translocate these back to the coral host for its nutrition,[49] clearly transforming the relationship from a neutral commensalism or one-sided exploitation to a reciprocally beneficial (mutualistic) symbiosis. Hosting the algae within their own cells and receiving the photosynthetic products directly from the intracellular symbionts makes for an efficient way for corals to use the energy in sunlight and contributes to the great ecological success of coral reefs.

The utilization in photosynthesis of host waste products is important because coral reefs, especially those in mid-ocean, occur in clear, blue water low in planktonic food – clear and blue because it has scant concentrations of the inorganic nutrients needed to support the growth of phytoplankton, which would cloud the water and turn it greener. Discovering these nutritional relationships in corals contributed greatly to the modern view of the zooxanthellae as recyclers of scarce nutrients. The coral host also feeds on relatively rare animal plankton and in turn fertilizes the zooxanthellae with its metabolic wastes, permitting the development of coral reefs as verdant oases in the blue desert of nutrient-poor tropical oceans.

Illumination of the dinoflagellate endosymbionts also enhances calcification by their coral host,[50] perhaps by supplying organic molecules and oxygen that stimulate the host animal's respiration and thereby provide energy needed for the calcification process, and perhaps some of the inorganic carbon (CO_2) that ultimately is incorporated into the calcium carbonate ($CaCO_3$) of the skeleton. The algae may also contribute to forming the organic matrix where the crystallization of $CaCO_3$ occurs. The algae might indirectly help to remove protons (hydrogen ions, H^+, the concentration of which is expressed in the pH scale, values below the neutral value of 7 being acidic and higher values being alkaline) produced in the host's metabolism, including the calcification reaction itself.[51] In so doing they would help control the pH, making the mineralization site more alkaline and more conducive to $CaCO_3$ deposition. To be sure, deep-sea corals lacking algal partners calcify, too, but at only one-tenth the rate of symbiotic corals in the light.

But to reap the harvest of their photosynthetic endosymbionts, corals must expose themselves to searing tropical sunlight, including its ultraviolet (UV) wavelengths, which have many detrimental effects, including sunburn and skin cancer in humans. So, why don't corals get sunburned?[52] In a shared metabolic enterprise,[53] corals and their algal symbionts jointly synthesize natural sunscreens called mycosporine-like amino acids (MAAs) that intercept the energetic UV radiation and dissipate it harmlessly as heat before it reaches sensitive cellular components.[54] Field and laboratory studies implicate UV radiation as the dose-dependent stimulus for corals to synthesize MAAs. Thus, concentrations of MAAs are greater in corals living in shallow water than those in deep water because UV levels are lower in deep water, following their attenuation by seawater. There has been considerable interest in using MAAs (particularly those produced by some seaweeds) in human UV protection and skin care.

Molecular oxygen, so abundantly produced photosynthetically by zooxanthellate corals, becomes a menace at high levels and, particularly in bright sunlight, promotes the formation of oxygen free radicals and other reactive oxygen species (ROS) injurious to cellular DNA, proteins and membranes. In humans, ROS are intimately involved in diverse pathologies that contribute to ageing. According to Steve Jones in *Coral: A Pessimist in Paradise* (2003), 'We cannot live without the gas for more than a few minutes but cannot live with it for more than a few decades.' Against such 'photooxidative stress', corals (both the host animal and the endosymbiotic algae) must also maintain biochemical defences, including DNA repair systems as well as antioxidant enzymes such as superoxide dismutase, catalase, various peroxidases and small-molecule antioxidants to remove ROS.[55] By keeping highly energetic UV radiation from reaching sensitive targets inside the cell, MAAs also help to avoid photooxidative stress. As we have seen, their robust defences against cellular insults may contribute to the slow rate of ageing in some corals, and it may not be coincidental that the longest-lived gold and black corals live deep in dark water.

The Colours of Corals, and Green Oases in Blue Deserts

Symbiotic corals have other defences against the double-edged sword of their need for solar radiation with its attendant dangers. Without light, there would be no colours. Throughout the rapturous descriptions of corals across the ages, their resplendent, even fluorescent, colours have enchanted and puzzled observers – as one author titled a paper, 'Why are there Bright Colors in Sessile Marine Invertebrates?' There has been no satisfactory answer to this question. One idea is that the luminous colours of corals act as lures to attract prey, but other factors also are involved.

In 1867, Baron Eugen von Ransonnet-Villez published his paintings of underwater scenes of coral reefs, sketched first in a diving bell. In one of these, *Submarine Rocks with Green Corals*, several species of corals clearly stand out from their shaded backgrounds, highlighted by their brighter green. Ransonnet marvelled that 'green itself is of an astonishing brilliancy and even becomes comparatively brighter at a remote distance.'[56] Some of this luminosity is attributable to the optics of seawater and the depth at which Ransonnet made his observations (see Chapter Four) and to the characteristics of human vision (the overall luminosity response, or perceived brightness, of our eyes' photoreceptors is greatest in the green wavelengths, and would be accentuated in the dim shadows), but there may be more to it (illus. 64).

Although it was first found in their relatives the sea anemones, corals provided some of the most spectacular examples of the phenomenon of fluorescence, in which specimens irradiated with short wavelengths of light (including invisible UV) transformed these and re-emitted longer wavelengths of visible light. The first scientific report, in 1944 by Siro Kawaguti studying corals in Palau (Micronesia), indicated green to be the most common colour of fluorescence, but he incorrectly ascribed it to the zooxanthellae. Eventually this colour was linked to green fluorescent protein (GFP; see illus. 40), discovered in a jellyfish, then tweaked molecularly to yield additional colours, and subsequently commercialized and applied widely in biomedicine 'to reveal the unseen'[57] workings inside cells. These discoveries and developments were recognized by the 2008 Nobel Prize in Chemistry award.

Others of Ransonnet's prints show additional natural colours of fluorescent coral pigments. Then, in 1956, René Catala in Noumea, New Caledonia (site of the world's second-largest coral reef system), catalogued the wondrous array of fluorescent colours emitted by living corals under UV lamps: green, orange, red, blue, beige and brown, and silver. A full-page colour photograph of Catala's fluorescent subjects in the popular U.S. weekly *Life* magazine in 1959 caught the imagination of the wider public.

Later that year Catala shipped living fluorescent corals for an exhibition in Antwerp, Belgium. He published his book *Carnival under the Sea* in 1964, including photographs of coral fluorescence and a preface by a member of the French Academy of Sciences, which read in part:

> In the darkness of an evening, turning the beam of an ultra-violet lamp upon the deep-sea corals, CATALA suddenly reveals an astonishing spectacle. As if touched by the wand of a fairy, all the polyps change their colors. The varied and beautiful hues adorning them by daylight vanish, and are replaced by a fairyland of glittering gems which dazzle the observer. Rubies, emeralds and topaz glow here and there at the tips of the tentacles, around the mouths, along the bodies of the flowering madrepores as their outlines become lost in the deep shadows of the tanks. (illus. 65)

About the same time, in Arthur C. Clarke's novel *Dolphin Island* (1963), written from personal experience for adolescent readers, a blue/UV flashlight became a 'magic wand' during a night-time dive on the Great Barrier Reef and caused corals 'to burst into fire, blazing with fluorescent blues and golds and greens in the darkness'.

The beautiful hues of corals (see illus. 63) and other anthozoans in daylight arise from diverse molecules such as carotenoids (familiar in orange, yellow and red vegetables, as well as autumn leaves) but especially chromoproteins, not all of which (notably blue and purple variants) are fluorescent. The fluorescent and non-fluorescent pigments are located only in the host animal's cells, not in the symbiotic algae. (At high density the algae impart their own green, olive and brown aspect to the surrounding coral tissues owing to their chlorophyll and accessory photopigments involved in photosynthesis.) A part of each chromoprotein molecule acts as the chromophore (Greek, 'colour-bearer') that is responsible for the perceived colour. Differences in the chromophores' structures (amino acid composition, sequence and configuration) account for their different colours. The fluorescent proteins (FPs) of anthozoans constitute a family of biochemically related molecules, evinced by their molecular sequences and resultant structures. Absorption of excitation wavelengths in the presence of O_2 causes a structural change in the chromophore (specifically, a fluorochrome) and its re-emission of light at longer wavelengths, which is seen as fluorescence.

Opposite page: 64 The brightness of some corals seen underwater is accentuated by green fluorescent protein. Impression of plate VII, 'Submarine Rocks with Green Corals', from Eugen von Ransonnet-Villez's book Sketches of the Inhabitants, Animal Life and Vegetation in the Lowlands and High Mountains of Ceylon, as Well as of the Submarine Scenery near the Coast, Taken in a Diving Bell *(1867). Below: 65 Colourful corals become more brilliant when they fluoresce under short-wavelength blue light. Paired photographs in daylight (left) and shortwave blue light (right).*

Such a striking phenomenon demanded a functional explanation. Despite more than a quarter of a century of research, there has been no consensus. Neither enhancement of photosynthesis by fluorescence under dim conditions nor photoprotection of the algae in bright sunlight has been definitively established as a role for coral fluorescent proteins.[58] Nor do FPs seem to be important as supplementary UV sunscreens, considering the much greater UV-absorption efficiency of MAAs, which unlike FPs are more concentrated at shallower depths where solar irradiance is greater,[59] and moreover have their biosynthesis and cellular concentrations greatly enhanced by exposure to solar UV radiation.[60]

Although a protective function and a mechanism remain elusive, there is a correlation between the concentration of FPs in the tissues of some corals and their resistance to bleaching (loss of their endosymbiotic algae).[61] GFP might protect the host by dissipating high-energy blue wavelengths. Others have suggested that FPs and non-fluorescent chromoproteins (CPs) act as supplemental antioxidants to help remove ROS such as superoxide radical and hydrogen peroxide.[62] The heat-resistant FPs may assume greater importance when antioxidant enzymes become compromised by the high seawater temperatures associated with coral bleaching. Under such stress, the pigmented algae depart

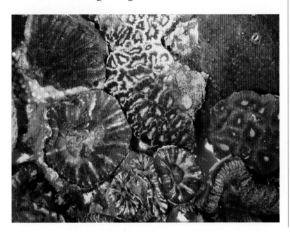
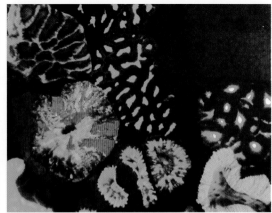

or are expelled by the host, leaving the bone-white skeleton visible through the host's FP- and CP-tinted translucent tissues, now without their most important source of nutrition and at risk of reduced growth or even death if the algae are not restored. Coral bleaching is one of the most striking manifestations of the warming temperatures that are a consequence of global industrialization (illus. 66).

The fashionable conceit of coral reefs, especially oceanic atolls, as colourful, idyllic oases in the blue desert of tropical oceans originated not from the knowledge of coral symbioses as recyclers of scarce nutrients, but rather with low coral islands as oceanic havens for colonizing seeds of future terrestrial vegetation. According to John Barrow, 'no sooner have the points of the coral rocks reached the surface, so as to form a barrier for the accretion of adventitious matter floating on the waves, and by its accumulation rise into an islet, than the seeds of the vegetable world burst into life.'[63]

In the same passage in *A Voyage to Terra Australis* where in 1802 he extolled the monument of the corals' wonderful labours, Matthew Flinders described the formation and greening of a coral cay:

> The new bank is not long in being visited by sea birds; salt plants take root upon it, and a soil begins to be formed; a cocoa nut, or the drupe of a pandanus is thrown on shore; land birds visit it and deposit the seeds of shrubs and trees; every high tide, and still more every gale, adds something to the bank; the form of an island is gradually assumed; and last of all comes man to take possession.

Opposite page: 66 In bleached staghorn corals devoid of symbiotic algae, the pigment pocilloporin in the coral host gives the colony tips their brilliant blue colour. Overleaf: 67 According to Percy R. Lowe in A Naturalist on Desert Islands *(1911), 'born of the depths of the sea, and raised but little above the waves, this bare reef became, through their [cast-adrift plants' and bird-borne seeds'] efforts, a little green oasis in the midst of the blue expanse of water.'*

Once inhabited by humans, who imported additional plants and animals, the soil improved through the continuous reworking of waste.

Percy R. Lowe restated the thesis in *A Naturalist on Desert Islands* (1911): 'born of the depths of the sea, and raised but little above the waves, this bare reef became, through their [cast-adrift plants and bird-borne seeds] efforts, *a little green oasis in the midst of the blue expanse of water*' (emphasis mine). The conceit was born (illus. 67). The role of recycling by the corals themselves was added by brothers Eugene P. and Howard T. Odum in their pioneering ecology textbook of the 1960s *Fundamentals of Ecology*, and later conveyed to a wider audience as an object, even moral, lesson by Eugene in a 1998 essay. There he revivified zoophytes: 'The idea that coral is an interdependent plant–animal cooperating complex or system was a radical notion at the time, but it is now accepted as one of the main reasons why coral reefs prosper in a nutrient-poor environment.'[64] Ironically, in an era of warming climate, it is coral bleaching – the loss of the algae and breakdown of this intimate association – that is killing corals and threatening reefs worldwide.

3 The Mythos, Menace and Melancholy of Corals

They are monsters, true, but monsters of beauty.
Umberto Eco, *The Island of the Day Before* (1995)

These under formings in the mind,
Banked corals which ascend from far,
But little heed men that they wind
Unseen, unheard – till lo, the reef –
The reef and breaker, wreck and grief.
Herman Melville, *Clarel* (1876)

In the ancient Greek and Hindu myths that describe the origin of precious red coral, the key variations turn on whether the very blood of a monster slain by a hero became coral, or whether the flowing blood tinted the vegetation otherwise petrified by the monster's force. For Salman Rushdie, such oft-told stories 'lose the specificity of their beginnings, but gain the purity of essences' – here, blood. Reef-building scleractinians are taxonomically and culturally distinct from red coral in their origin. In Polynesia, the coral polyp was the first living being to emerge from primordial chaos, and corals in turn gave rise to humans and gods. Some Pacific heroes and deities remained associated with corals and reefs, assuming their form and solidity as a divine attribute or post-mortem monument. Where radiant red coral was blood, skeletal scleractinians were bone.

Opposite page: 68 According to the French caption to this engraving in Metamorphoses d'Ovide en rondeau *(1676), 'When Perseus had freed Andromeda and killed the monster, he placed Medusa's head, which had served as his shield, on certain grasses, which were changed into coral by its touch.'*

Myths of Corals
In his *Metamorphoses*, Ovid (43 BCE–17 CE) related Andromeda's rescue from the sea monster Cetus by Perseus. Having killed the Gorgon Medusa whose glance turned people to stone, for safekeeping the hero laid her serpent-coiffed head with its petrifying visage on a bed of seaweeds and twigs, which in later versions were turned into stone by the Gorgon's blood. But Ovid used the word *vim* (strength, force), not *cruorem* or *sanguinem* (blood), for what the twigs absorbed from Medusa's head, and had the twigs hardening at its touch (*tactus*). Charles Martin offered a modern translation faithful to the original:

> . . . before the hero washes his serpent-slaying hands, he carefully constructs a little nest there on the beach, of some soft leaves with seaweed strewn upon them, and there he rests Medusa's snake-fringed head, lest she be damaged by the beach's gravel.
> Thirsty fresh twigs, still living, still absorbent, soak up the monster's force, and at its touch rigidify through every branch and leaf. Astounded sea nymphs try experiments on other twigs and get the

same results; delighted, they toss them back into the sea as seeds to propagate this new species!

Coral today shows the same properties; its branches harden when exposed to air, and what was – in the water – a spry twig becomes a rock when lifted out of it.[1] (illus. 68)

In the absence of blood from the original version, it likely was the same force emanating from Medusa's visage – the *gorgoneion* – notoriously capable of turning men to stone (illus. 69) that also petrified the twigs, leaves and seaweeds where her snake-fringed head rested. Here the twigs rigidify at the monster's 'touch', which for ancient readers would have a visual connotation in an age when seeing was a form of touching at a distance. Still, Ovid did cast the vegetation as *bibula* (drinking or absorbent), so perhaps it was the *life force*, represented by flowing blood, that they soaked up and that he meant as the transformative agent.

Ovid further softened living coral: whereas it was a twig (*vimen*) in Book 4, in Book 15 it became a herb or grass (*herba*) beneath the waves and only hardened when exposed to air:

> So too with coral, which waves like
>   a grass underwater,
>   but being exposed to the air turns instantly
>   rigid.

Such suppleness suggests that Ovid may have been referring here to gorgonian corals – those elongated sea whips or reticulated sea fans that are relatives of the sturdier red coral – and whose underwater undulations feature in films of coral reefs, particularly those they typify in the Caribbean.

Opposite page: 69 The petrifying visage of Medusa, by Carlo Parlati II, late 1980s, Japanese cerasuolo coral, diamonds, pearls, 18K gold, malachite and marble.

Variants of the myth occurred in the Greek lapidary texts of the earliest Graeco-Roman period. Collected by French scholars, these texts described the properties of various minerals, as coral was then considered. First,

> They say it owes its existence to Perseus who, having decapitated Medusa from behind, washes himself of the taint of the murder in the sea, and to do this places the severed head on the grass; the blood drips on the plants that become tinted purple, while the daughters of Oceanus – swift breezes – with their breath make the blood congeal. It becomes rigid stone but keeps its vegetal shape.[2]

This version of the myth conveys that coral is what today would be called a laminate – here, plants tinted by Medusa's blood which, after coating them, hardens on them as a crust. Precious red coral indeed mirrors such a laminated, composite structure, but inversely, with a solid, red, calcareous central axis surrounded by layers of more diffusely calcified soft tissue of variable hue. In another telling of this version, the roots of the sea-plants absorb the blood, which fills the plants, consistent with coral's red axial core.

The second variant recalls the later passage in Ovid (Book 15), where he does not mention coral's origin from the death of Medusa, only that soft, herbal coral turns hard in air. We hear it lyrically in Ann Wroe's *Orpheus: The Song of Life* (2012):

> He sang with special fascination of the *kouralion*, or coral, born as a plant in the sea-deeps, so delicate, so light, paddling to the surface like a small creature and then, tired of swimming, turning into stone as it rested under the clear 'sky-air' on the shore.

It is this version also that is attributed to the elusive Orpheus in his poem 'Lithica' (but probably written by a pseudo-Orpheus in the fourth century CE and gathered with other lapidary texts). Orpheus, who may have lived a generation before Homer, was an 'author more cited than preserved by classical antiquity'.³

The origin of red coral became explicitly bloodier with its reinvention in the Renaissance. Michael Cole, who also had noticed that Medusa's blood did not feature in the text of this episode in Ovid's *Metamorphoses*, quoted the earliest (fourteenth-century) translation of the work in Italian, by Giovanni Bonsignore,⁴ which separated the petrifaction of the twigs from their being coloured by the blood, as in the 'laminated corals' of the lapidary texts.

Cole presented his insight in a treatise on Benvenuto Cellini's bronze sculpture *Perseus with the Head of Medusa*, cast between 1548 and 1552 (illus. 70). Cellini may have known Ovid from Bonsignore's translation but in envisioning the sculpture omitted the twigs and transformed blood directly into coral, mindful of the decapitation – unavoidably bloody – of Medusa. It was Cole's insight that here lay the critical transformation in Renaissance eyes: the Gorgon not only petrified people, but she herself – her life's blood – became rigidified.

This realization in turn crystallized, for Cole, Cellini's originally enigmatic mention of casting '*due gorgoni*' (two Gorgons), which he elsewhere specified to be the '*gorgoni* of the neck and of the head of Medusa' – the *blood* pouring from these two parts of her body. Cole solved the puzzle after reading in a mid-nineteenth-century Italian dictionary: 'the plural *gorgonii* names not Gorgons as such, but rather a particular family of polyps – *gorgonii* are corals.' With Cellini, it seems, originated the conceit that the very blood of the exsanguinated Medusa metamorphosed into coral – so that corals were born directly of her death.

Earlier, Andrea Mantegna's painting *Madonna della Vittoria* (Madonna of the Victory, 1496) had featured a superb red coral suspended above the Virgin and Child. The colony's shape and colour prefigured the flowing blood of Christ's Passion and his Resurrection (and perhaps, too, being a *cornuto* symbol warding off evil) (illus. 71, 72 and 73). Coral, and indeed the polyp itself, retained this association with blood. An anatomical cast in resin of the human cardiac circulation, with its branching or rooted form, recalls a colony of red coral (illus. 73). In Samuel Johnson's *A Dictionary of the English Language* (1755), 'polypus . . . is likewise applied to a tough concretion of grumous [clotted] blood in the heart and arteries.' In a pensive reversal of blood becoming coral, the contemporary artist Daniel Arnoul shaped and polished Mediterranean red coral to represent droplets of blood trickling from a pricked finger in his 1989 ebony sculpture *Doigt de Mars* (illus. 74).

The story of Medusa and her bloody death no doubt has ancient antecedents. Basilio Liverino reminds us in *Red Coral: Jewel of the Sea* (1989) that the earliest hunters would have associated the loss of blood from their prey's wounds with its gradual dying. 'Perhaps instinctively, man realized that without that "redness" life could not exist; and instinct again suggested that he wore on his own body something of the same colour which could substitute for the other if necessary.'

The association of blood and life inevitably ramified in myth and religion. G. Evelyn Hutchinson (the polymath best known for his work in aquatic sciences), in musing on coral, offered that:

> We may guess that behind the succession of Uranos, Kronos and Zeus lies a series of divine kings, mutilated or killed by their successors, and that the generative power of water and the abundance of the sea was

Opposite: 70 Benvenuto Cellini, Perseus with the Head of Medusa, *1548–52, bronze.*

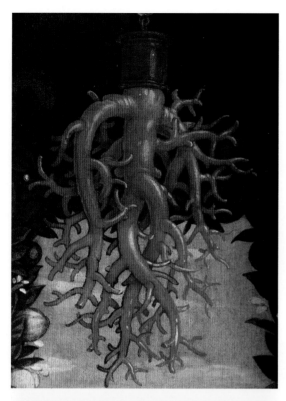

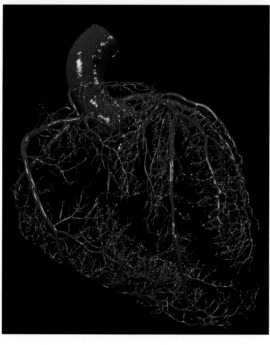

Opposite page: 71 Andrea Mantegna, Madonna and Child, *known as* Madonna of the Victory (Adored by Gian-Franceso II Gonzaga), *1496, oil on canvas. Top left: 72 detail of red coral from the painting; bottom left: 73 resin cast of the human cardiac circulation, showing its resemblance to a colony of red coral. Below: 74 Daniel Arnoul,* Finger of Mars, *1989, ebony, red Mediterranean coral, agate, gold.*

somehow supposed to depend on such bloody practices . . . through some violence to a god . . . who through this hurt was a benefactor of mankind. It is hard to avoid the suspicion that the color of coral . . . implies his blood.[5]

Giovanni Tescione, in his book *Il corallo nella storia e nell'arte* (Coral in History and Art; 1965) quotes an unnamed twentieth-century Neapolitan poet who encapsulated the image:

Coral produces a flower of pain,
distills and makes of itself the essence
    of light
in scintillating clotted blood.

The ancient association of Mediterranean red coral with blood was continued in a ritual whereby the Oba of Benin in Nigeria restored the spiritual power of his coral adornments by bathing them in the blood of a human sacrifice (nowadays a cow's blood is used). And in Tabarka, Tunisia, even in modern times a black cow was sacrificed aboard a coral-fishing boat and its blood sprinkled onto the boat, the nets and the sea to assure safe navigation and a bountiful harvest, as well as to encourage the coral to thrive.[6]

Giorgio Vasari incorporated Cellini's bloody imagery of the metamorphosis in his own painting *Perseus Freeing Andromeda* (1570), where Medusa's head rests upright on its severed neck, directly on the shore and not on an interposed nest of twigs and seaweeds, and the ramifying rivulets of blood stain the sea sanguine (illus. 75). The sea nymphs collect the branches of coral emerging prolifically from this confluence. Under Vasari's direction, Francesco Morandini, his long-time assistant, contributed to the decoration of the *studiolo* of Grand Duke Francesco de' Medici, including the painting *Water*, on the vault of the dressing room, where Tritons and Nereids, elaborately posed as in Vasari, display their iconic attributes of pearls, shells and red coral.

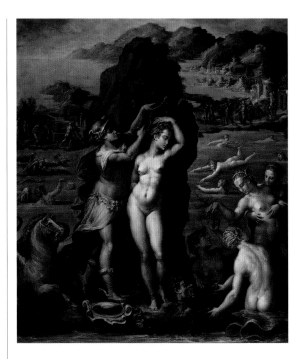

Jacopo Zucchi's *The Treasures of the Sea* (c. 1585) (illus. 76) virtually duplicates the topography, lighting and posed personages of Vasari's painting. Luxuriant coral immediately draws the eye in both paintings (the central figure in Zucchi's holds branching examples of the red and rarer white forms – recalling blood *and* flesh or bone – of Mediterranean *Corallium*). In both cases patches of red in the corals and garments (and even in a parrot's head in Zucchi's work) successively draw the eye to different parts of each painting and hold the elements together, simultaneously complementing the blue of the sea and sky. But Zucchi puts coral to another use: the painting is also known as *Allegory of the Discovery of the New World*, and the assembled exotica of coral, pearls and nacreous shells, as well as foreign animals and non-European humans, are symbolic of the wealth and wonders lying beneath

Above: 75 Georgio Vasari, Perseus Freeing Andromeda, *1570, oil on slate. Opposite page: 76 Jacopo Zucchi,* The Treasures of the Sea, *c. 1585, oil on copper.*

and beyond the seas. In Celeste Olalquiaga's reading of this scene, 'The ocean floor has been subordinated to all kinds of colonial fantasies.'[7]

Others subsequently would paint Perseus' rescue of Andromeda, but the link to coral disappeared from their representations of the maiden chained to the rock. It was not until Sébastien Bourdon's mid-seventeenth-century painting *The Liberation of Andromeda* (illus. 77) that the legend of coral was resurrected. Here, Bourdon put the *gorgoneion* on Perseus' reflective shield, where sea nymphs offer up plants that are turned into branches of red coral, not by Medusa's blood but by her visage.

Pierre Mignard (court painter to Louis xiv and director of the Académie Royale) reprised the theme in *Perseus Liberating Andromeda* of 1679 (illus. 78). Here, Medusa's serpent-coiffed head lies (face down!) on a bed of seaweed, although there is no flowing blood, but instead realistically rendered red coral branches (such as were to be seen at the time in curiosity cabinets and as the handles of cutlery) emanating, almost coursing, directly from among her locks of hair and snakes.

The origin of red coral so treasured in India appears in the Hindu legend celebrating Lord Vishnu's defeat of the demon king Bali. In one telling, Indra, the lord of heaven, was being overthrown by demons led by Bali, and to save humankind Indra appealed to Vishnu. Thus, on behalf of Indra, Vamana (the dwarf Brahmin avatar of Vishnu) petitioned Bali for a refuge of as much land as he could enclose in three steps. Bali agreed, whereupon Vishnu assumed his

Opposite page: 77 Sébastien Bourdon, The Liberation of Andromeda, *1650, painted canvas. Above: 78 and detail (below) Pierre Mignard,* Perseus Liberating Andromeda, *1679, oil on canvas.*

godly proportions and in two steps encompassed heaven and earth. With no place for Vishnu's third step, Bali offered his head for Vishnu's foot, which instantly crushed Bali and turned his body to gems. Drops of Bali's blood gave rise to rubies, whereas his blood that flowed to the sea became branched red coral.[8]

Alternatively, Indra's thunderbolt cut Bali's body into pieces that became gems, including red coral from his tongue.[9] The centrality of gems may be a recent embellishment and figures prominently in the importance of gemstones in

Vedic astrology.[10] Precious stones, including coral, also appear widely in older Buddhist traditions.

There are scant accounts from ancient China or Japan about the origin of red Mediterranean coral, although it was eagerly sought in trade because of its auspicious colour. In China, it represented longevity and new beginnings, including official promotion. Wei Yongwu's eighth-century poem 'Yong Shanhu' placed coral's birth on Mount Penglai in the sea east of China, in the mythical land of gods and immortals. Red coral is prized in Tibet as one of the Buddha's sacred stones, symbolizing the life force and being used in personal ornamentation to promote success, as well as in medicine and religious art.[11]

In Japanese mythology, the dragon god Ryūjin lived in an undersea castle made of red and white coral.[12] In an old folktale from the early ninth century, a sea turtle (another symbol of longevity) took the young fisherman Urashima Tarō to Ryūjū Castle, where he saw trees of red coral in an exotic underseascape (illus. 79). Tarō also visited Mount Hōrai (the Japanese name for Mount Penglai), where he saw a palace surrounded by jewels and coral.

Precious red coral occurs in the popular Japanese folktale of Momotarō the Peach Boy.[13] When Momotarō was a young man, a gang of ogres was victimizing the local villagers. Momotarō set off to the ogres' island to engage them, taking with him several animal helpers. After the vanquished ogres promised to mend their ways, Momotarō and his companions returned home bearing the ogres' treasure (illus. 80). The coral tree made its first appearance as part of the treasure in a printed version of the story in 1785 and again in 1805, a time when exotic, valuable red coral (previously restricted to the

Opposite page: 79 Painting on the ceiling of the Nishidomari Tenmangu shrine in Otsuki, western Kochi Prefecture (the early centre of Japanese fishing for precious corals), showing Urashima Tarō returning from Ryūjū Castle with precious corals and a casket containing a surprise. Above: 80 Harunobu Hishikawa, A Scene of the Folk Tale Momotarō, Depicting Momotarō's Return with Treasures after Slaying Daemon, *1890, print on paper, triptych.*

shoguns, high-ranking samurai, and wealthy merchants) saw increasing use in accessories and decoration, and knowledge of it became widespread even among commoners. As in other cultures it had medicinal uses, including defence against poison, and for people of Edo-period Japan (1603–1868) as 'a talisman against evil spirits, a magnet for good fortune, a rare and exotic treasure'.

A rather different coral may have featured in the ancient Sumerian epic poem *Gilgamesh*, which chronicled that hero-king of Uruk (*c.* 2700 BCE). Towards the end of the poem, Gilgamesh learns that if he can obtain a certain treelike, thorny plant that lives beneath the sea, it will rejuvenate him and give him eternal youth. Tying heavy stones to his feet, Gilgamesh sinks to the bottom of the sea (perhaps at Bahrain, in the Arabian/Persian Gulf), collects the plant and returns to the surface with bloodied hands but looking forward to testing the plant on an old man before trying it himself. Elizabeth During Caspers intuited that the marvellous undersea plant (likened to a 'dog rose' or a 'buckthorn') might be an antipatharian coral, noted for its spines that indeed prick one's hand and for its restorative, even aphrodisiac, qualities (illus. 81).[14]

Across the Indian Ocean, in northeastern Arnhem Land, on the northern coast of Australia, washed by the Gulf of Carpentaria and the Sea of Arafura, the Dreaming of the Warramiri clan of the Yolngu people includes the creator being Ngulwardo, the king of the sea. This being is embodied as the coral reef, that is, the creation of scleractinian corals, atop submarine bedrock. In the 'laboratory' of the reef, Ngulwardo transforms itself into various sea creatures. The whale is the foremost of Warramiri totemic sea creatures, and its bones are made of Ngulwardo's coral.[15]

Still farther eastward, Polynesian peoples dispersed across the Pacific, often living on coral islands (drawn up from the depths by an

ancestral fisherman) that became the solid seat of their cultures in the vastness of the ocean. In a Hawaiian creation myth, the Kumulipo, the tiny scleractinian coral 'insect' or polyp was the first life to emerge from a shadowy chaos, giving rise to 'perforated' coral colonies and heads that formed reefs. Coral, then, was the progenitor of all living things. The Polynesian creation included the 'conception of an evolutionary succession':[16] worms and molluscs followed the zoophytes, and the decaying bodies of marine organisms – the slime – eventually rose above sea level, where terrestrial organisms successively appeared on that foundation, culminating in the appearance of gods and humans. A prayer to the creator god Kāne recited his names, which included several forms of coral.[17] Elsewhere, north of Fiji, the first human was formed directly from vapour emitted by coral rocks.

In addition to Hawaiian gods and demigods, there were subordinate *akua* such as Opuhala, responsible for giving birth to corals and their reefs, which were among her body forms.[18] It was Opuhala who provided huge hooks made of coral (or, variously, shell) that the hero Maui used to fish up the islands of the Pacific. Opuhala is associated not only with coral but also canoe bailers.[19] This pairing may not be coincidental when we consider both the story of Cook's crew manning the pumps to save the *Endeavour* after its holing on a coral reef, and Tahitian folklore in which Pua Tu Tahi, or Coral Rock Standing Alone, was an undersea monster that rose to the surface and turned into jagged rock to tear the keels out of fishing boats.[20]

A Hawaiian genealogical chant written in 1700 by the poet Keaulumoku on the same outline as the Kumulipo creation myth was recited to Captain James Cook when he was welcomed as the god Lono at Kealakekua in 1779. Martha Warren Beckwith reminded us that 'coral was the first stone in the foundation of the earth.'[21] The deposed Liliuokalani, last Queen of Hawaii, completed an English translation of the chant in 1897 during her imprisonment and specifically recognized the coral 'insect' as the creator of the 'perforated coral'.[22]

Coral Icons and Amulets

Kenneth Oakley, an archaeologist, established an earliest date for the use of 'perforated coral' (fossilized scleractinians) when he described a flake of fractured chert composed of an Upper Jurassic specimen. This artefact was associated with Acheulian toolmakers in Swanscombe, Devon (*c.* 200,000 BCE). The transversely broken or polished surface of such a coral showing the radially symmetrical corallites may 'recall the heavens to the untutored mind', because in British folklore such rock was called 'starrystone'.[23] The importation of the chert from a distant site, despite its being less readily worked than local flint for tools, suggested a symbolic rather than a practical importance: 'Did [the starry] patterns have symbolic value even to the earliest members of our species . . . ?' 'Is it conceivable that even 0.2 Ma [million years] ago they called to mind the night sky?' Two other fossils – a gastropod and another 'madrepore' or starrystone – from a site associated with late Neanderthal Mousterian industry (35,000 BCE) 'were found close to two nodules of pyrite and a perfectly spherical stone "bola". Were these four objects part of the equipment of a Mousterian hunter depending for success on charms in addition to weapons, or the belongings of a Mousterian "sorcerer"?'[24]

Opposite page: 81 Thorny antipatharian corals: Louis Roule, Antipathaires et cérianthaires, planche III: Leiopathes grimaldii et Aphanipathes [Antipathes] erinaceus. Résultats des campagnes scientifiques accomplies sur son yacht par Albert Ier, Prince Souverain de Monaco, fascicule XXX, Déscription des antipathaires et cérianthaires recueillis par S.A.S. le Prince de Monaco dans l'Atlantique Nord (1886–1902), *1905, lithograph.*

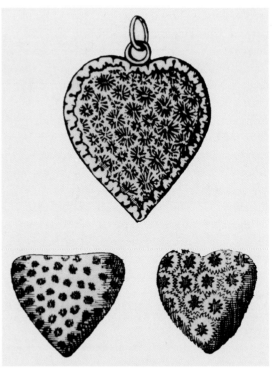

Left: 82 Upper Palaeolithic (Magdalenian, c. 10,000 BCE) pendant of fossilized coral. Above: 83 Verschreiherzen, or 'hearts of charm' pendants. Fig. 8 in Alexander von Schouppé, Episodes of Coral Research History up to the 18th Century *(1993).*

Fossilized solitary corals were pierced to form beads at a Cro-Magnon Aurignacian culture site in western Russia.[25] Half a world away, in a Malaysian cave inhabited by humans 40,000 years ago, a hand axe was fashioned by breaking the head from a domed fossilized corallum that presented a similar starry pattern, evidence of magical power.[26]

Personal ornamentation abounded in the European Magdalenian culture of about 15,000 years ago. One such ornament is a pendant made from a piece of fossilized coral that combines the heavenly sidereal pattern with a strikingly human, perhaps pregnant 'Venus' form (illus. 82).

Fast-forward to sixteenth-century Europe, when the Swiss zoologist Conrad Gesner figured fossilized corals called 'Astroites' or in German 'Sternstein', congruent with the British starrystones. These lucky stars were thought to bring good fortune and were used both as amulets in battle and, cut into the shape of a heart, worn in Bavaria and Austria as apotropaic pendants (illus. 83),[27] consistent with the Paracelsian belief that the stars were linked to human organs and diseases.

More recently, the Petoskey stone (the state stone of Michigan) has a celestial link, but with no supernatural aura. Fossilized colonial rugose coral *Hexagonaria percarinata* from Devonian reefs is polished to highlight the radiating pattern of its corallites and used in trinkets and keepsakes. It was named after an eighteenth-century Ottawa Native American: when the rising sun fell auspiciously on his newborn face his father called him Petosegay, meaning 'rising sun' or 'rays of dawn'. The polished Petoskey stone is used as a 'worry stone', rubbed for relaxation or relief of anxiety. President Barack Obama had one on his desk in the White House (illus. 84).

Belief in the 'evil eye' has been among the universal superstitions, in part because vision was the most mysterious of the senses and therefore was tied to magic.[28] By casting a baleful glance, a malicious *jettatore* was able to produce a malevolent effect – *fascinare* is the Latin verb – in the receiving person. Thus, diverse cultures developed magical defences against the evil of 'fascination', as detailed by F. T. Elworthy in *The Evil Eye* (1895). Most effective was an amulet – a *fascinum* – worn around the neck of women and especially children (deemed the most susceptible) to divert the gaze of the *jettatore* seeking to cast the evil eye. Such a charm would be doubly effective if it also confused and mortified the *jettatore*, distracting his evil intention. A phallus filled the prophylactic bill (illus. 85).

In classical Rome, parents hung a phallic amulet around the neck of infants and children as a defence against bewitchment. The culturally widespread representation of the phallus was coloured red, and sometimes the amulet was made not of the more common bronze or other metal, but of coral. A device of bright red coral would have the dual advantage not only of being eye-catching and diverting, but also of embodying a material that itself protected children from harm, as Pliny had written. He told, too, that Indian soothsayers and diviners held coral to be a sacred protection against all ills and spells, an attribute espoused by Zoroaster as well. Although Pliny specified only a half-dozen or so medical uses for it, coral collectively in world cultures was a panacea, appearing in diverse pharmacopoeias for use in cordials, electuaries (medicinal pastes with honey), powders and syrups, and its salutary effects extended beyond the medical, improving the fertility not only of women but of fields, as a protection against storms, hail and lightning, and against scorpion stings and rabies in dogs.

The Roman phallic amulet (called the *res turpicula*, 'indecent thing') often was a dual charm, with the phallus at one end and at the

Left: 84 President Barack Obama strokes a Petoskey stone as he talks on the telephone in the Oval Office, 6 December 2012.

other the *mano fico* ('fig hand') – a gesture in which the thumb is clenched between the index and third fingers and represents the vulva. The gesture is still used in southern Italy and other cultures both as a protective sign and in a dismissive, obscene context. Another gesture, the *cornuto*, in which the index and little fingers are extended as horns and the thumb and other fingers are bent towards the palm, also invokes protection from evil. Representations of the *mano fico* and the *cornuto* are commonly carved from coral and used as amulets, especially in southern Italy and Sicily. The sanitized phallus shows up also as a simple curved horn in coral suspended from the neck (illus. 86, 87 and 88).

Left: 85 Graeco-Roman phallic pendant, c. 3rd–1st century BCE, red coral and gold. An eye-catching phallus carved from red coral and worn as an amulet would divert the gaze of a jettatore seeking to cast the evil eye on the wearer. Above: 86, 87 and 88 Carved amulets of red coral (left to right): the mano fico *and* cornuto *signs, and a horn.*

In medieval Italy, the branch of coral itself in its manifestation of the Tree of Life was the amulet of choice for children, appearing in religious portrayals of the Madonna and the infant Jesus. Coral also saw religious uses in vestments, crucifixes, chalices and diverse religious carvings, and became all the fashion as paternoster beads and rosaries, although some Catholic orders saw these as too luxurious. In fourteenth- and fifteenth-century paintings, the Christ Child often had a coral branch suspended from his neck, as part of a Franciscan humanizing of a more compassionate Divine Child. A branch of coral remained as a children's charm well into the Renaissance period. The red of coral was a multifaceted symbol of Christian piety, particularly among the royal and wealthy, and its use increased during the Renaissance. In some cases, the coral was paired with another amulet – the cross – as in Jacopo di Mino's *Madonna col Bambino* (Madonna and Child) of 1342 (illus. 89).

Revisiting Dangerous Corals

The twin themes of destruction and death date to the earliest writings about coral reefs. Pliny the Elder repeated stories that undersea forests of Indian Ocean corals tore the rudders off Roman ships. Among the early first-person accounts of the danger posed by these reefs was that by François Pyrard de Laval, shipwrecked in the Seychelles in 1602:

> Being in the middest of an Atollon, you shall see about you a great ledge of Rockes which impale and defend the Iles, against the impetuousnesse of the Sea. But it is a very fearefull thing even to the most couragious to approach to this ledge, and see the waves come afarre off and break furiously on every side . . . The surge or billow is greater than a House, and as white as Cotton: . . . a very white wall, especially when the Sea is loftie . . . The depths of the Sea are generally very keene and sharpe Rockes which hurt them wonderfully that goe into it.[29]

A quarter-century after Pyrard's experience, the Dutch East Indiaman *Batavia* on her maiden voyage in 1629 ran up on a coral reef under full sail at Houtman Abrohols in the Indian Ocean off Western Australia. The archipelago had been charted barely a decade earlier by Frederik de Houtman, adapting the place name from a Portuguese mariner's warning, *abri vossos olhos*

89 Jacopo di Mino del Pellicciaio, Madonna col Bambino *(Madonna and Child)*, detail, 1342, tempera on panel, Church of San Martino, Sarteano, Italy. The child's necklace shows the unusual pairing of a protective branch of red coral and a crucifix.
90 (above) This detail was sketched by G. Evelyn Hutchinson for his essay 'The Enchanted Voyage: A Study of the Effects of the Ocean on Some Aspects of Human Culture' (1995), Journal of Marine Research.

– keep your eyes open. The aftermath of the wreck saw mutiny and mass murder, as retold by Mike Dash using historical first-hand accounts in *Batavia's Graveyard* (2001).

A century and a half later, on the other side of the continent, the sharp-eyed Lieutenant James Cook had been threading the *Endeavour* through the maze of reefs that Banks would call a 'Labyrinth of Shoals', where despite the leadsman's soundings indicating deep water ahead on a moonlit high tide, the ship ran up and stuck fast on what is now called Endeavour Reef. The rapid shoaling and steepness (particularly on the seaward or windward side of reefs – the sailor's dreaded lee shore, being on the leeward side of a ship sailing there) figured in many other accounts of these hazards to mariners. The crew and the gentlemen worked desperately to lighten the ship and manned the pumps. They deployed an anchor and a cable around a capstan to kedge her off the reef at the next high tide, and over the next two months made sufficient repairs to continue the voyage (illus. 91).

A particular hazard of living coral rock was its sharpness, graphically fictionalized by Dash, where the *Batavia* 'howled as shards of coral gouged their way along her sides . . . her bottom grating ominously against the coral'. Soon, survivors of the wreck 'found themselves surrounded on three sides by raging surf and claws of coral . . . listening to the awful grating of the hull'. Cook described at first hand the cuts in *Endeavour*'s hull, where 'scarce a splinter was to be seen but the whole was cut away as if it had

91 HMS Endeavour *stuck fast on a reef near Cape Tribulation at 11 p.m. on 11 June 1770. After refloating the ship, the crew subsequently beached her at the mouth of the Endeavour River to make repairs and then continued the voyage. From Andrew Garran,* Picturesque Atlas of Australasia, *vol. 1 (1886).*

been done by the hands of Man with a blunt edge tool.' As Banks journalized, 'we were upon sunken coral rocks, the most dreadful of all others on account of their sharp points and grinding quality which cut through a ships bottom almost immediately.'[30] These properties of corals were exploited in Pacific island cultures for use as tools for boring, grinding and sanding.

Following the lead of an Irish sea captain, Captain Jules Dumont d'Urville aboard the new *Astrolabe* sailed to the Vanikoro islands in 1828 in search of the expedition of Jean-François de Galaup, comte de Lapérouse, lost in a 1788 disaster that would colour France's maritime aspirations and expeditions for years afterward. With Dumont d'Urville were the surgeon-naturalists Quoy and Gaimard, who were continuing their study of corals. Quoy's account of the loss of Lapérouse's ships on the barrier reef surrounding the islands[31] was sensationalized by Jules Michelet in his popular book *The Sea* (1864), invoking a 'terrible lee shore' with its 'peaked and jagged rocks . . . as cutting as razors'.

The compelling details of their grounding given by Cook and Banks were later appropriated by the prolific writer of high-seas adventures Patrick O'Brian in *The Thirteen-Gun Salute* (1989), set during the Napoleonic Wars. Jules Verne in *Twenty Thousand Leagues under the Sea* (1870)[32] put his characters in similarly dire straits, redolent of Cook but more explicitly Dumont d'Urville, who had run up on a coral reef in the Torres Strait between far northern Australia and New Guinea in 1840 'and nearly went down with all hands'.

Peter Matthiessen set his modern adventure novel *Far Tortuga* (1975) in the Cayman Islands and remote cays off Honduras, in the Caribbean. The tragedy began as the author cast a mood of foreboding, with dark coral heads gathering like underwater storm clouds while a turtle schooner's crew moved cautiously among channels in the reef. It ended with the ship's break-up and sinking,

92 A diver explores the steep slope at Tydeman Reef, Great Barrier Reef.

'encircled by white wraiths of reefs', on a dark and stormy night on the Misteriosa Bank.

New dangers have surfaced in the modern era of adventurous ecotourism. The rapidly shoaling reef front, dangerous for Cook and other mariners, presents the inverse problem for scuba divers. According to the free-diver James Hamilton-Paterson in his chapter 'Reefs and Seeing' in *The Great Deep* (1992), 'the seaward cliff of a reef face reaches to within inches of the surface and may plunge 3,000 feet in a precipitous slope, its initial slope marked by a steady change in colors and lifeforms.' For Arthur C. Clarke's young protagonist in *Dolphin Island* (1963), 'Sooner or later, he would have to follow Mick down that blue, mysterious slope.' Beyond that, Charlie Veron warns scuba divers, there is the ever-present threat of depth, for clear-water reefs that plunge down into deep ocean can be deceptive, and all too often become a fatal attraction for an enthusiastic diver who wants to get just a little more out of the trip of a lifetime . . . With the water crystal-clear and the reef sloping steeply, at 50 meters the face looks much as it did at 10 meters – a death trap for divers, for it gives no sense of depth . . .

. . . The profusion of life everywhere gives divers pause – to stop, to rest, to think, and to marvel about the life around them . . . These can only be fleeting thoughts, however, for divers must never give in to the reef's siren call.[33] (illus. 92)

Death and Pathos

Within the poetic and literary spheres, there was always an undercurrent of death and pathos associated with hidden coral reefs. For generations of mariners, who saw death by drowning as removing all hope of immortality, there was the dreaded danger of shipwreck: recall the opening canto of *Clarel* (1876) by that literary seaman Herman Melville, where hidden danger became manifest:

> Unseen, unheard – till lo, the reef –
> The reef and breaker, wreck and grief.

A seaman himself, Philip Freneau (the 'Poet of the American Revolution') wrote 'The Hurricane' while 'at Sea, in a Heavy Gale' off the coast of Jamaica in 1784, mindful that in the deep, dead sailors 'On coral beds unpitied sleep'. The Australian poet J. J. Donnelly, informed by the Great Barrier Reef, in a dark and watery 'Sea-Wraith' wrote of a shipwreck victim,

> . . . a sailor craving the balm of sleep
> in the coral gardens of the deep.

Their mysterious ambiguity, and their living hidden in the horrifying depths of the sea, long linked corals and the polyps themselves with death. The 'polypus' was a recurring symbol in the mythological poetry of William Blake (who knew the nature of coral), where in Night IV of *The Four Zoas* (c. 1796–1802),

> The Corse [corpse] of Albion lay on the Rock; the sea of Time & Space
> Beat round the Rock in mighty waves, & as a Polypus
> That vegetates beneath the Sea, the limbs of Man vegetated
> In monstrous forms of Death, a Human polypus of Death.

The recognition that reefs and atolls were built by animals kept the focus on the builders – the coral architects – especially in the poetic imagination, which was often touched by death. Surely the greatest literary monument to the builders themselves is James Montgomery's epic poem *The Pelican Island* (1828), explicitly inspired 'by a passage in Captain Flinders's *Voyage to Terra Australis*'. Montgomery saw the reef as a

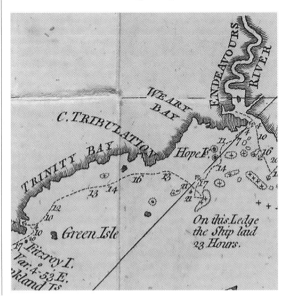

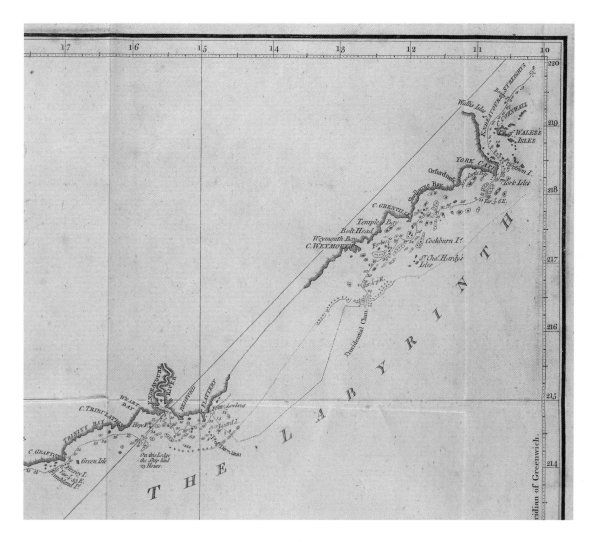

'stony eminence' inhabiting the zone 'where light and darkness meet in spectral gloom, midway between the height and depth of the ocean', and a 'marvellous structure climbing tow'rds the

93 and 94 Details of James Cook's chart showing THE LABYRINTH *and Endeavour's grounding site near what Cook named Cape Tribulation because 'here begun all our troubles.' The crew repaired the ship at what is now the Endeavour River (opposite). The distance from Cape Tribulation to Cape York spans 5 degrees of latitude (366 nautical miles or 678 km).*

day' – a 'labyrinth' of 'imperishable masonry' built instinctively from 'slime' by the 'petrific touch' of 'worms' that 'radiated like stars' and 'bred and died' and built their own 'tombs' atop their ancestors – all in a process 'which out of water brought forth solid rock'.

Despite its lyrical treatment of the natural history of corals and reefs, Montgomery's poem was nevertheless deeply tinged with death, where the ephemeral architects, in 'dying upward', were building their own 'temple of oblivion' that included 'mausoleums' and 'sepulchers' (the last word also was used in 1810 by Robert Southey in

The Curse of Kehama, for the title of Canto XVI, containing his floral description of a coral reef). While questioning whether it is 'within the sphere of science to question the poet',[34] James Dana (a geologist on the U.S. Exploring Expedition of 1838–42, later a professor at Yale, who presented the first evidence for Darwin's subsidence theory) opined that Montgomery's image arose from his too-exuberant anthropomorphizing and misunderstanding the nature of the serial stacking of the corallites. There, each polyp moves upward as it closes off each successive corallite cup in secreting a new one, and is not itself entombed in the sealed cell.

On his chart of the barrier reefs off North Queensland, Cook labelled them not a mere 'maze' but 'THE LABYRINTH', with its deadly connotation of a monster and a hero who cheated death there. Given that those reefs extend across more than 5 degrees of latitude (678 km/366 nm), Cook also needed the longer label to span them! (illus. 94)

Simultaneously with the publication of Montgomery's epic, Dumont d'Urville sailed from Tasmania to the fateful Vanikoro islands.[35] Although the natives told Dr Gaimard 'that the sea holds all the bones of the men who were shipwrecked', the party recovered no human remains from Lapérouse's expedition. In Dumont d'Urville's account of the 'terrible reefs of Vanikoro' and the finding of the vestiges of Lapérouse's ships, he referred to the debris as 'pathetic'. After visiting the coral-encrusted remains and recovering some of the artefacts, the ship's party set about building (out of coral rock) a memorial to the lost ships (illus. 95). Dumont d'Urville first called it a 'mausoleum', but because it contained no bodies later changed the word to 'cenotaph'. With sickness rampant in his crew, and himself racked with fever, he worried

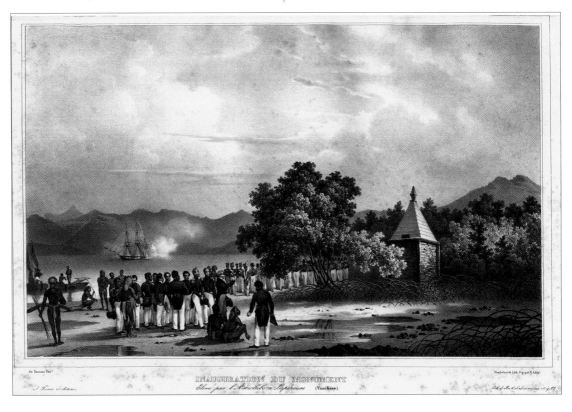

A. FLOTSAM.—WRECK OF NEW GUINEA MISSION SCHOONER "HARRIER."

Opposite page: 95 With a salute from the ship's cannon, the officers and men of the new Astrolabe *dedicate a monument of coral rock to the La pérouse expedition, watched by local islanders and within sight of 'the terrible reefs of Vanikoro'. Lithograph of a painting by Louis Auguste de Sainson (1828?), in Jules Dumont d'Urville,* Voyage de la corvette l'Astrolabe: Historique Atlas III *(1830–34), plate 187. Above: 96 Wreck of the schooner* Harrier *on the Great Barrier Reef near Cooktown. Photograph by W. Saville-Kent in* The Great Barrier Reef of Australia *(1893), plate XXX. Overleaf: 97 Eugen von Ransonnet-Villez,* Ceylon unterseeische Küstenlandschaft (Ceylon Coastal Undersea Landscape), *n.d., after 1865, oil on canvas. Elements of the scene were taken from Ransonnet's paintings that were published as lithographs in his book of his travels and underwater observations in Ceylon.*

that the structure might become their own memorial. Jules Verne (who had read Dumont d'Urville's account) reprised the tale in his chapter 'Vanikoro' in *Twenty Thousand Leagues under the Sea*, where from the *Nautilus*, and not a small surface craft, Professor Arronax 'gaped at this pitiful wreckage'.

Joseph Beete Jukes, sailing on HMS *Fly* in the Torres Strait, climbed on a recent wreck on a reef being pounded by furious breakers, a melancholic experience:

> Now and then some higher wave than usual would burst against the bows of the wreck . . . and travelling along her sides would lash the rudder backwards and forwards with a slow creaking groan, as if the old ship complained of the protracted agony she endured.[36]
> (illus. 96)

Charles Darwin's *Beagle* diary entry regarding coral architects included this funereal context:

> In time the central land would sink beneath the level of the sea and disappear, but the coral would have completed its circular wall. Should we not then have a Lagoon Island? Under this view, we must look at a Lagoon Isd as a monument raised by myriads of tiny architects, to mark the spot where a former land lies buried in the depths of the ocean.

And recall Lyell's letter to Herschel regarding Darwin's subsidence theory, where atolls were 'the last efforts of drowning continents to lift their heads above water'. Our contemporary writer Adam Gopnik is succinct and evocative: 'A coral reef is just a funeral wreath around the tip of a defunct mountain.'[37] To T. A. Stephenson (a marine ecologist on the 1928–9 Great Barrier Reef Expedition), seen from a great height, 'Large and solid though these reefs are, they seem . . . to float on the sea like wraiths.'[38] Funereal reefs are wreaths and wraiths.

The disquieting, sepulchral guise of the outlandish growth forms of corals was analogized as a conceit of period writing. Edward John Trelawny, Percy Bysshe Shelley's 'friend and champion', repeatedly rewrote his account of the poet's 1822 drowning and seaside cremation over the course of five succeeding decades, and the Romantic biographer Richard Holmes saw those embellishments 'accumulating more and more baroque details, like some sinister biographical coral reef'.[39]

The association of corals with human death, interment and petrifaction pervaded the nineteenth century. Baron Eugen von Ransonnet-Villez, the first person to sketch views of coral reefs while underwater (see Chapter Four), included a human skull in one superb scene, which became a *vanitas* painting congruent with the nineteenth-century imagination of the ocean depths (illus.

97). Jules Rengade presented the Coral Sea not only as a swallower of ships, but also where the dead 'were forever entombed in its somber caverns'.[40] The image of entombment was used twice by Jules Verne in *Twenty Thousand Leagues under the Sea*, first in the chapter 'Vanikoro', when Arronax and Nemo view the vestiges of Lapérouse's ships: '"Ah, what a beautiful death for a sailor," the captain was saying. "A tomb of coral is a peaceful tomb. Heaven grant me that my companions and I rest in no other!"'

The image is more graphic in the chapter 'The Realm of Coral', where Nemo intones, 'We dig the graves, and we entrust to the polyps the task of sealing away our dead for eternity.'[41] Verne's prose became more memorable with the accompanying drawing by Alphonse de Neuville, 'Everyone Knelt in an Attitude of Prayer', showing

Opposite page: 98 Henri Hildibrand's engraving of Alphonse de Neuville's drawing 'Everyone Knelt in an Attitude of Prayer', in Jules Verne, Vingt milles lieues sous les mers *(20,000* Leagues under the Sea*; 1871). Above: 99 Filming the undersea funeral in Walt Disney's* 20,000 Leagues under the Sea *(1954). Overleaf: 100 For Gustave Flaubert, coral colonies (*Dendrophyllia ramea*) had massive arms, which for Hans Christian Andersen were slimy and had worm-like fingers to seize both hapless humans and other beings in the deep.*

the crew of the *Nautilus* kneeling around an open tomb with a rough-hewn coral cross at its head and towering coral formations looming behind (illus. 98). Stuart Paton included the scene in his film of 1916, based loosely on the novel, in which the Williamson brothers pioneered the use of underwater cinematography. More popularly, the 1954 Walt Disney filmic version of the novel likewise lingered in the successful scene (illus. 99).

Verne's verbal, and later the filmic, image of pickaxes piercing the stony coral substratum to dig the grave recalled the commentaries by Pyrard de Laval, Cook and Banks, Flinders, and Quoy about the reciprocal lacerating danger of razor-edged reefs. As coral reefs became increasingly visited, so did excruciating encounters with them. Painfully sharp coral became embedded in the fictional characters of Melville in his autobiographical *Omoo* (1847) and Umberto Eco in *The Island of the Day Before* (1994). It was a small step for the villain in the James Bond film *For Your Eyes Only* (1981) to turn luxuriant lithophytes into malevolent madrepores – instruments of death by a thousand cuts – and to tow Bond (and his obligatory female companion) across a reef behind a speeding boat, scattering acroporid debris in its bloody wake.

In his surreal film *Deux cent milles lieues ous les mers, ou le cauchemar du pêcheur* (*Under the Seas*; 1907), Georges Méliès had already placed maleficent corals among the beings that threatened the dreaming fisherman Yves after he emerged from his submarine. Here, however, Méliès contravened the conception of corals as stationary stone: their jagged branches were articulated and mobile, nearly ensnaring Yves as he tried to smell their *fleurs du corail*.

Cory Doctorow, too, in his twenty-first-century online short story 'I, Row-Boat',[42] embellished the simile of the coral snare as a weapon that sentient, mobile corals used to trap scuba divers in a cage, 'a thicket of calcified arms', taking revenge for human devastation of coral reefs. Both Méliès' and Doctorow's imagery recalls Georges Cuvier's use of the term *piège* (snare or trap) to describe the hazard presented by the stony trunks of corals intertwined into rocks and reefs.[43] For Hans Christian Andersen's 'The Little Mermaid' (1837), who lived in her father's undersea palace made of coral, seashells and amber, the danger lay in the grasping vegetation of a witch's wood, where many coral conceits were entangled in a few short sentences. There,

all the trees and flowers were polypi, half animals and half plants; they looked like serpents with a hundred heads growing out of the ground. The branches were long slimy arms, with fingers like flexible worms, moving limb after limb from the root to the top. All that could be reached in the sea they seized upon, and held fast, so that it never escaped from their clutches . . . She saw that each held in its grasp something it had seized with its numerous little arms, as if they were iron bands. The white skeletons of human beings who had perished at sea . . . even a little mermaid, whom they had caught and strangled; and this seemed the most shocking of all to the little princess.[44] (illus. 100)

Calcifying and regenerating corals frequently have been linked not only with human death but also transmutation and immortality. One of the earliest literary expressions of this transition was that of Alonso, by Shakespeare in *The Tempest* (*c.* 1610), with his now-clichéd 'Ariel's Song' in Act I, Scene ii:

Full fathom five thy father lies;
Of his bones are coral made;
Those are pearls that were his eyes;
Nothing of him that doth fade,
But doth suffer a sea-change
Into something rich and strange.

What was Shakespeare's inspiration for the post-mortem metamorphosis of bones into coral? One suggestion was that Alonso's sea change was his resurrection, long symbolized religiously by red coral. The pairing of the precious imperishables, pearls and red coral (associated since antiquity as emblems of the oceanic Nereids), would cement Alonso's undersea persistence after his metamorphosis

and was perhaps inspired by costumes in Inigo Jones and Samuel Daniel's masque *Tethys, Queen of Nymphs and Rivers* (1610).[45]

But precious *red* coral did not have an artistic or literary tradition associated with human skeletons. Certainly Shakespeare was aware of the myth of Medusa and the origin of red coral from reading Ovid (on whose *Metamorphoses* he drew heavily[46]) in his Latin studies as a schoolboy or from Arthur Golding's English translation of 1567, which may have seeded the idea of coral and human metamorphosis. Simply seeing a scleractinian skeleton or white variant of precious coral and recognizing the osseous aspect, Shakespeare may have invented the sea change of bones into reef coral, even richer and stranger than twigs into precious coral, and without the chromatic dissonance of *red* coral as metamorphosed human bone (illus. 101 and 102).

Melville later would use such scleractinian skeletal imagery in his novel *Mardi* (1849), in retelling the myth associated with the origin of the coral reef around the tropical island setting. The god Upi, an archer, slew an evil giant whose 'remains petrified into white ribs of coral' could be seen by glancing over the gunwale of a boat passing above. Such a transformation also occurred in the legend of the muscular twelfth-century hero Ono (Hono'ura) in the Marquesas (French Polynesia). Ono's body became a coral reef after he was decapitated at Atuona (later the site of Paul Gauguin's grave).

Elsewhere in *Mardi*, Melville commingled human bones and coral while describing a Pacific island king: 'But what sways in his hand? A scepter, similar to those likenesses of scepters, imbedded among the corals at his feet. A polished thigh-bone . . . For to emphasize his intention utterly to rule, Marjora himself had selected this emblem of dominion over mankind.' And the royal dwelling

was chiefly distinguished by its pavement, where, according to the strange customs of

Opposite page: 101 Edmund Dulac, 'Full fathom five thy father lies; Of his bones are coral made', *painted for* Shakespeare's Comedy of The Tempest with Illustrations by Edmund Dulac *(1908), watercolour, gouache, and pen and brown ink on paper. Above: 102 Mark Dion,* Bone Coral (The Phantom Museum), *2011, papier-mâché and glass.*

the isle, were inlaid the reputed skeletons of Donjalolo's sires; each surrounded by a mosaic of corals, – red, white, and black . . . 'Thy own skeleton, thou thyself dost carry with thee, through this mortal life, and aye would view it, but for kind nature's screen, thou art death alive . . .' And over the Coral Kings, Babbalanja paced in profound meditation.

Melville recycled the simile of human bone and coral in a funereal verse (set in a mausoleum in Jerusalem) of his epic poem *Clarel* of 1876, and garlanded it with floral imagery as well:

Hard by, as chanced, he once beheld
Bones like sea corals; one bleached skull
A vase vined round and beautiful
With flowers; felt, with bated breath
The floral revelry over death.

In his mystery thriller *Bones of Coral* (1991), James W. Hall tells us that the original Spanish name of Key West, Florida, surrounded by coral reefs, was Cayo Hueso, or Bone Key, after the human remains of an earlier battle – the skeletons of the native inhabitants scattered on the white coral sand. Callum Roberts in *The Ocean of Life* (2012) describes the thirty-year aftermath of the 1982 mass bleaching of corals in the eastern Pacific caused by El Niño:

By the end, the Galapagos reefs were all but dead . . . and structures built over thousands of years have now crumbled to rubble and dust . . . It was a scene of devastation. The

seabed was strewn with chalky fragments, like a field of bones in the aftermath of a massacre.[47]

Glyn Philpot conjoined such meditations on death and coral in his painting *Under the Sea*, begun in Venice in 1914 on the eve of the First World War (illus. 103). Here an armless statue of a naked female form in marble (metamorphosed limestone) lies on the seafloor, surrounded (as was the drowned Alonso in the Dulac watercolour of six years earlier) by diverse zoophytes. But in Philpot's work, they are mostly gorgonians, sea anemones and sponges – some real, some apparently fanciful – and a smattering of hard corals, not all of which occur in the Mediterranean or Adriatic Sea. The scene is not a natural one from the floor of those seas, so Philpot may

Opposite page: 103 Glyn Warren Philpot, Under the Sea, *1914–18, oil on canvas. Above: 104 An assemblage of zoophytes and lithophytes (including the deep-sea scleractinian* Loph[oh]elia pertusa*), dredged aboard the research vessel* Travailleur *by Alphonse Milne-Edwards (Henri's son).*

have been stylizing and exaggerating the forms and colours that he saw in public aquariums, in natural history books or in museum collections[48] (a similar array occurs in the contemporaneous collections of Prince Albert I of Monaco, in the Institut Océanographique in Paris) (illus. 104).

Note that in Philpot's painting, the statue's submergence must have been recent: it is not yet encrusted by corals or other marine organisms that opportunistically colonize hard artefacts, which thereby become substrata for new reefs. Philpot's statue is vivified by the works of the contemporary artist Jason deCaires Taylor, who places human forms of cast concrete in reef areas, where encrusting organisms colonize them and change their visage and texture over time (illus. 105).

Philpot's statue may symbolize the demise of classical Greek or Roman civilization, or the danger to contemporary European civilization as war loomed.[49] The pale, armless statue recalls the Venus de Milo, the importance of which was recognized in 1820 by none other than the young Dumont d'Urville, who, immediately after it had been excavated on the Aegean island of Milos, arranged its purchase for France. The classical statue and the aura of loss in Philpot's subaqueous scene also summon up Atlantis: recall

Lyell's comment about drowned continents and Gopnik's analogy of atoll reefs as funeral wreaths above them.

The solitary, drowned figure among the corals conjures up Richard Garnett's poem 'Where Corals Lie' (1859), orchestrated by Edward Elgar in 1899. Here, the possibly suicidal poet drowns, drawn by the alluring beauty of 'the land where corals lie':

> The deeps have music soft and low
> When winds awake the airy spry,
> It lures me, lures me on to go
> And see the land where corals lie.
>
> By mount and mead, by lawn and rill,
> When night is deep, and moon is high,
> That music seeks and finds me still,
> And tells me where the corals lie.
>
> Yes, press my eyelids close, 'tis well,
> But far the rapid fancies fly
> To rolling worlds of wave and shell,
> And all the land where corals lie.
>
> Thy lips are like a sunset glow,
> Thy smile is like a morning sky,
> Yet leave me, leave me, let me go
> And see the land where corals lie.

The nineteenth-century physician C. B. Klunzinger, too, felt this allure when viewing the coral reefs of the Red Sea:

> We feel drawn downwards as it were by a mysterious power towards these objects, apparently so near, yet rendered by the foreign element so distant and unattainable, and we gaze dreamily into the depths, sunk in nameless feelings and dim impressions.[50]

Opposite page: 105 Jason deCaires Taylor, Vicissitudes, *2007, cement cast.*

Corals and reefs have attracted their fair share of romantically inclined, introspective, escapist and even melancholic biologists. Alfred Goldsborough Mayor, once an assistant to Alexander Agassiz, pioneered the experimental biology of corals, having donned a diving suit to explore the Great Barrier Reef, and later founded the first tropical marine laboratory in the United States at Dry Tortugas off Florida in 1904. According to his obituary in *Science* in 1922, Mayor was 'of a distinctly artistic and poetic temperament'. It was this Mayor who, in his beautifully self-illustrated volumes *Medusae of the World* (1910), wrote, 'Love, not logic, impels the naturalist to his work.'[51]

Iain McCalman quoted from Mayor's unpublished journal, written during a visit to Turtle Reef, off Cooktown, during Agassiz's 1896 Great Barrier Reef expedition:

> As one gazes down through the deep turquoise depths and sees the lovely play of color that the sunbeams revel in among the branches of the coral forest where shadow vies with sunshine to enchant the beholder's eye, it seems another world far from this earthly realm of ours. A place far removed from the struggle of the upper world, and where sorrow is unknown and life goes on forever in listless, languid happiness and beauty.[52]

Later, back in Florida, Mayor wrote to his wife: 'I . . . look far down into the recesses of the coral caverns where the cool deep shadows invite one to plunge beneath our prosaic world into the brilliant enchanted one below.'[53]

4 Conjuring Corals

> those absolute bouquets formed in the depths by the alcyonaria, the madrepores. Here the inanimate is so close to the animate that the imagination is free to play infinitely with these apparently mineral forms . . . a cluster drawn from a petrifying fountain.
> André Breton, *L'Amour fou* (1937)

How were pensive people to treat the manifold manifestations of corals? Cultivated from the collective human imagination, these arborescent entities gradually were trimmed with contemporaneous cultural conceits. Floral corals were arranged into bouquets and flowerbeds or vegetable gardens, and the redness of precious corals made them the Tree of Life or symbols of Christ's Crucifixion and Resurrection. Calcareous corals also were crystals that formed spontaneously, as ice did, solid but delicate and transient, too. Corals' rocky permanence saw naturalists and princes construct cabinets to contain their curiousness. Redly redolent or starkly skeletal, corals assumed human anatomy as flesh and blood or bone. Artists and scientists painted isolated atolls as halcyon idylls, emblematic of harmonious nature, apart from yet alluring to human experience. The sonorous majesty of surf on the reef, and corals' literary and poetic history, inspired musical compositions that took listeners into the not-so-silent world beneath the waves.

Parterres of Corals

Nineteenth-century seafarers and naturalists experienced and wrote, often ruminatively, occasionally rapturously, not only of the menace and melancholy but especially of the beauty inherent in the equivocality of coral reefs,

Opposite page: 106 Edmund Dulac, Birth of the Pearl, *c. 1919–20, in Leonard Rosenthal,* The Kingdom of the Pearl *(1920).*

sometimes informed by the unfolding knowledge about them. Matthew Flinders, commander of the *Investigator*, extended Cook's charting of 'The Labyrinth' of reefs, naming it a 'Great Barrier'. On 9 October 1802,

> I went upon the reef with a party of the gentlemen; and the water being very clear round the edges, a new creation, as it was to us, but imitative of the old, was there presented to our view. We had wheat sheaves, mushrooms, stags horns, cabbage leaves, and a variety of other forms, glowing under water with vivid tints of every shade betwixt green, purple, brown, and white; equalling in beauty and excelling in grandeur the most favourite *parterre* of the curious florist . . . but whilst contemplating the richness of the scene, we could not long forget with what destruction it was pregnant.[1]

Here Flinders was perpetuating the similitude of corals and terrestrial plants, even though by then corals were well known to be animals. But the botanical image, deep-rooted over centuries, was robust, and it was only natural to use familiar things to describe strange and novel ones, because people map what they know onto what they don't. Flinders, as commander of the vessel, was responsible for its safety and so concluded a passage otherwise replete with romanticized allure and grandeur with a reminder of the danger of reefs, particularly because at the time he was trying to find a safe way out of the Labyrinth.

Elsewhere Flinders matter-of-factly described the construction of coral reefs by 'animalcules', as had John Barrow, a Fellow of the Royal Society on an early voyage to Cochinchina, now part of Vietnam. But the naturalist Barrow also enthused: 'How wonderful, how inconceivable, that such stupendous fabrics should rise into existence from the silent, but incessant and almost imperceptible, labours of such insignificant worms!'[2] Had Barrow read Erasmus Darwin's epic poem *The Botanic Garden* (1791), in which reefs were 'Where living rocks of worm-built coral breathe'? In his 'Philosophical Note' to this line, Darwin wrote of the 'Immense and dangerous rocks built by swarms of coral insects'[3] and used Cook's description of them as perpendicular walls. A few lines earlier, the elder Darwin touched all of coral's animal, floral and mineral bases:

> Feed the live petals of her insect-flowers,
> Her shell-wrack gardens, and her sea-fan bowers;
> With ores and gems adorn her coral cell,
> And drop a pearl in every gaping shell.
> (illus. 106)

Similarly, William Blake's familiar print from around 1805 seated the figure of Isaac Newton underwater, on what appears to be the colourful living rock of a coral reef that

107 William Blake, Newton, *1804–5, colour print finished in ink and watercolour on paper.*

includes sea anemones (illus. 107). Blake likely knew of the multifarious nature of corals: in his youthful poem 'To the Muses', written in 1783, he placed the Muses 'beneath the bosom of the sea', wandering on 'crystal rocks . . . in many a coral grove'.[4]

In 1810 Robert Southey, one of the English Lake Poets and later Poet Laureate, published his epic *The Curse of Kehama*, set in a submarine Indian city that he described in Canto XVI, 'The Ancient Sepulchres', with its 'coral bowers, grots of madrepores, and arborets of jointed stone', their 'living flowers in open blossom spread, stretch'd like green anthers many a seeking head', and 'trees of the deep, and shrubs and fruits and flowers wherewith the Sea-Nymphs love their locks to braid'. Here came together Romantic poetry and its familiarity with natural history and perhaps Renaissance painting, but with corals retaining their old domestic botanical and stony duality.

This twin stony and vegetable nature of corals recurs in other texts, such as Herman Melville's description of the 'coral grove' in Papeete harbour in *Omoo* (1847):

> Down in these waters, as transparent as air, you see coral plants of every hue and shape imaginable: – antlers, tufts of azure, waving reeds like stalks of grain, and pale green buds and mosses. In some places, you look through prickly branches down to a snow-white floor of sand, sprouting with flinty bulbs . . . (illus. 108)

In 'The Coral Grove' (1863), the poem for which he is most remembered, James Gates Percival brought together both agricultural and sylvan botany, with a measure of marine biology and geology. As the oceans became increasingly domesticated in the age of the Victorian aquarium, allowing wider experience with marine organisms, the French writer Arthur Mangin would chuckle knowingly in *The Mysteries of the Ocean* (1864): 'Marine nature seems to amuse itself by giving to the inferior animals forms in imitation of those of the terrestrial vegetables.'

The floral image of polyps as blossoms – Marsigli's *fleurs du corail* – continued well after they had become animals. The expansion and contraction of polyps was a simile that blossomed famously when Philip Henry Gosse applied it to sea anemones in his *Actinologia Britannica* (1860):

> While the glare of day is upon them, they are often chary of displaying their blossomed beauties; but an hour of darkness will often suffice to overcome the reluctance of the coyest . . . If you would make sure of seeing it in all the gorgeousness of its magnificent bloom, visit your tank with a candle an hour or two after nightfall.[5]

In 1893, the Cuban-born French poet José-Maria de Heredia would visit the warm depths of a coral reef at dawn to recombine coral animals in full bloom with flowering plants in the first verse of 'Le Récif de corail' (The Coral Reef):

> The submarine sun, mysterious aurora,
> Lights the forest of Abyssinian corals
> Which intermingle, in the depths of their
>   warm pools
> The blossomed animal and the living flora.

In his book *Physiologie des animaux marins* (1938), Paul Portier, the co-discoverer of anaphylaxis in response to cnidarian stings, wrote of his experience in the sunlit calm of the Great Barrier Reef. There, 'thanks to the perfect transparency of the water, we could follow the . . . corals "in flower".'

To the late nineteenth-century poet Jules Laforgue, entering the grotto of the Berlin aquarium seemed an immersion in his own inner world, and yet also a fusion with the universe

and a dilation of the self, which he described as the expanding or blossoming of a coral, '*se madréporiser*'.[6] John Cowper Powys likewise linked coral and imaginative mental expansion in his novel *Porius* (1951): 'His thoughts renewed themselves, expanded their cramped tendrils, glowed richly coral-red in their ancient sun-spawned freshness' (illus. 109).

Reefs built by corals offered habitat for a myriad other animals that gradually were appreciated. In 1847, the same year that Melville published *Omoo*, Joseph Beete Jukes, the naturalist and geologist, published his account of the surveying voyage of the *Fly*, and like Charles Darwin not long before, 'had hitherto been rather disappointed by the aspect of coral reefs, so far as beauty was concerned; and though very wonderful, I had not seen in them much to admire'.[7] But as had Darwin's, Jukes's attitude

Above: 108 Chris Garofalo, Corallum Desertus, *c. 2005, glazed porcelain. These ceramic chimeras comprise characteristics of corals. Some recall Melville's lagoon and its 'snow-white floor of sand, sprouting with flinty bulbs', where for Jules Michelet in* The Sea *(1864) 'you have the strong assemblage of aloes and cactus'. Opposite page: 109 Expanding polyps of* Corallum rubrum *extending their tentacles. Overleaf: 110 In Joseph Beete Jukes's horticultural description of reef corals,* Explanaria *(=* Turbinaria*) appeared as 'a great submarine cabbage garden', as in this photograph taken in Kiribati.*

changed one day while exploring the lee side of an outer reef,

> where every coral was in full life and luxuriance. Smooth round masses of maeandrina and astraea were contrasted

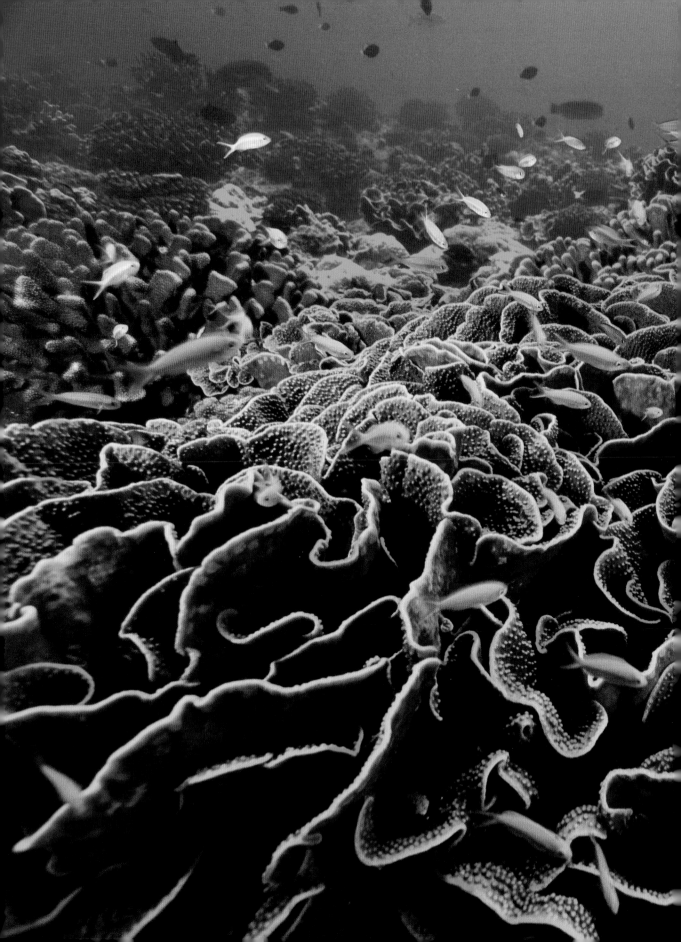

with delicate leaf-like and cup-shaped expansions of explanaria, and with an infinite variety of branching madreporae and seriatoporae . . . Their colours were unrivalled – vivid greens, contrasting with more sober browns and yellow, mingled with rich shades of purple, from pale pink to deep blue . . . In among the branches of the corals, like birds among trees, floated many beautiful fish, radiant with metallic greens or crimsons, or fantastically banded with black and yellow stripes.[8]

Thus Jukes, a romantic (who, like his contemporary Darwin, had studied at Cambridge for the clergy but was attracted instead by natural history, especially geology), was even more effusive than the ship's captain Flinders in specifying the types of corals and subtler in describing them and the fishes that flitted among their branches. But Jukes could not relinquish the persistent practical horticulture: a few pages later he saw the patches of grass-green coral (probably a foliaceous *Turbinaria*) as 'a great submarine cabbage garden' (illus. 110).

Alfred Russel Wallace, in describing the magnificent corals in the harbour at Ambon in *The Malay Archipelago* (1869), wrote again of colourful fish moving among 'these animal forests', but did not make the avian connection as Jukes had done. Jules Verne would use the simile of 'fish with fluttering fins . . . like flocks of birds' amid coral arborizations in his chapter 'The Realm of Coral' in *Twenty Thousand Leagues under the Sea* (illus. 111).

111 Nami ni chidori *(Wave and Plover), Japanese ornamental hairpin of carved tortoiseshell and pink coral.*

Matthias Jakob Schleiden lifted Jukes's description in its entirety to serve as the core of his own account of a coral reef and its denizens for his chapter 'The Sea and its Inhabitants' in the second (1853) edition of the English translation of his book of popular lectures, *The Plant: A Biography*. Jukes's metallic fish among the coral branches became for Schleiden 'hummingbirds of the ocean that play and flutter around the coral flowers', melding their seemingly metallic plumage and ability to hover in place above the flower-animals.

But the simile of floating fishes as hovering hummingbirds did not originate with Schleiden. C. B. Klunzinger, in his own description of Red Sea coral reefs in *Upper Egypt: Its People and Products* (1878), quoted the original source – the German naturalist and microscopist Christian Gottfried Ehrenberg's 1832 treatise on the same reefs, where: 'As the hummingbirds of the American hemisphere play around the flowers of tropical plants, so do small fishes, beautifully coloured with gold, silver, purple and azure, and hardly a few inches in size and never larger, play around the flower-like coral animals.'[9]

Ernst Haeckel, too, in *Arabische Korallen* of 1876 alighted on the hummingbird analogy and swelled the splendour of the setting to exceed that of the mythical garden of the Hesperides. In *The Island of the Day Before* (1995), Umberto Eco's effusive similes became parodies of published descriptions of such coral gardens, overflowing with vegetables, draped with fabrics and aflutter with enamelled hummingbirds.

Crystalline Corals

Ernst Haeckel saw the continuity of life from organism to atom and blurred the distinction between the organic and the inorganic. He hoped for a 'crystallography of organic forms' in his quest for the organizing principle of animals during their development, one attribute of the 'crystal soul' that was the title of Haeckel's final book and which was foreshadowed in his earlier illustrations of radiolarian and coral skeletons in *Kunstformen der Natur* (Art Forms in Nature; 1899–1904) (illus. 112).

This mineral, crystalline vision of coral had accreted for centuries, as in Flaubert's *The Temptation of Saint Anthony*, where some aragonite crystals – *fleurs de fer* or *flos ferri* – found in iron deposits resembled coral colonies draped as textured tapestries. Certain gypsum and anhydrite crystals (calcium sulphate, not calcium carbonate as in coral skeletons) such as ram's horns and desert roses also resemble coral colonies. Following the alchemists, whose work he studied, Robert Boyle (1627–1691), founder of modern chemistry and the Royal Society, had experimented with growing such colourful crystals that resembled corals or 'lovely Trees composed of Roote and Branches' by precipitating copper and cobalt salts.

André Breton, too, would compare coral and crystal in his Surrealist novel *L'Amour fou* (Mad Love; 1937), where 'there could be no higher artistic teaching than that of the crystal,'[10] which he saw as the perfect example of 'creation, spontaneous action'. He went on to point out the impermanence of living coral, constantly formed and destroyed in the sea. Amid mineral and floral imagery, he juxtaposed a photograph of magnified crystals of rock salt with one by the pioneering undersea cinematographer J. E. Williamson of what Breton called 'the treasure bridge of the Australian Great Barrier' (actually a reef structure in the Bahamas,[11] evident from the Caribbean elkhorn coral dominating the photograph) (illus. 113).

Breton crystallized corals,

> those absolute bouquets formed in the depths by the alcyonaria, the madrepores. Here the inanimate is so close to the animate that the imagination is free to play infinitely with these apparently mineral forms . . . a cluster drawn from a petrifying fountain

Opposite page: 112 Ernst Haeckel, in Kunstformen der Natur *(Art Forms in Nature; 1899–1904) emphasized the crystalline nature of scleractinian skeletons by enlarging the individual corallites to the size of coral colonies, which have their own different morphologies and polyp sizes (colonies at centre, top to bottom:* L. pertusa, M. meandrites filograna *and* Acropora humilis*). Above: 113 For André Breton, corals such as those in J. E. Williamson's photograph from c. 1934 were 'rock crystal towers with their summit in the sky and their feet in fog'.*

. . . [A] whole garden appears . . . the very substance of flowers, the fans of frost.[12]

Presenting mineralized corals as 'fans of frost' and 'clusters drawn from a petrifying fountain' was a masterstroke to visualize at once their form, crystalline beauty and impermanence. Such a vision was not unprecedented: the eighth-century Chinese writer Zhong Ziling already had likened the hardening of a coral tree to ice forming.[13]

Rivalling Breton's evocative text for its inventiveness and finely crafted beauty are the contemporary sculptures by Daniel Arnoul, who often combines red coral, crystals, other minerals, and metals. His *Self-portrait* (*c.* 1989) includes a medieval village of silver built as such villages were, to follow the terrain (here, in miniature, of a geode), surmounted by a rock crystal obelisk (as in Egypt, representing the sun's rays) and a blood-red coral as the Tree of Life. Red coral also occurs in a grotto, as it typically does in nature. The whole assemblage emanates from the partly submerged, insensate bronze head of the artist imagining it (illus. 115).

Coral Idyll

Coral reefs and idyllic atolls are the production both of polyps and human grey matter. Voyages of discovery had fixed tropical coral islands with their exotic vegetation and inhabitants in the European imagination, for example Louis-Antoine de Bougainville's 'happy island of Nouvelle Cythère', populated by Jean-Jacques Rousseau's uncorrupted 'homme naturel'. In their narrative of a scientific sojourn on an atoll in the Carolines, *Coral Island* (1958), the ecologists Marston Bates and Donald Abbott succinctly summarized generations of descriptions that often were tinged by colonialism: 'the word "atoll" carries with it the connotations of warm sun, of white beaches backed by graceful palms, of friendly dark-skinned people, and of an idyllic way of life.' Such a vision, in miniature, with a raised round bank of coral sand and a single palm tree, often with a hirsute shipwrecked figure or two, has by now become a caricature of the tropical desert isle.

The idyllic nineteenth-century image of coral atolls appeared in Darwin's diary from the *Beagle* expedition on 1 April 1836:

> As in the sky here & there a white cloud affords a pleasing contrast, so in the lagoon dark bands of living Coral are seen through the emerald green water.
> – Looking at every one & especially a smaller islet, it is impossible not to admire the great elegant manner in which the young & full grown Cocoa-nut trees, without destroying each others symmetry, mingle together into one wood: the beach of glittering white Calcareous sand, forms the border to these fairy spots.[14] (illus. 114)

Elements of Darwin's description – the dazzling

Below: 114 Ernest Griset, untitled triptych, 1871, watercolour on paper. The painting of an atoll had been intended as a birthday gift for Charles Darwin from his admiring neighbour, John Lubbock, and lying offshore in the background of the middle panel is what appears to be the HMS *Beagle. Opposite page: 115 Daniel Arnoul,* Self-portrait, *1989, geode, rock crystal, various minerals, red coral, silver, patinated copper, bronze and meteorite.*

whiteness of the sandy beach sprouting coconut palms, the brilliant hues of the water within and outside the lagoon, and the accents of a white cloud and foaming surf – recurred in travel narratives and literature throughout the nineteenth century, including the writings of Jukes, Ballantyne, Schleiden, Gosse and Melville.

Henri Matisse, visiting Tahiti in May 1930, echoed all and romanticized the familiar scene, analysing the colours of the reef and even experimenting with his perception of light above and under water:

> The lagoon . . . overlooked by a delicate white cloud . . . The elegant coconut palms with their locks rustled by the trade winds that blow endlessly, accompany the murmur of the open sea on the reef against the tranquil, undisturbed water of the lagoon . . . the colour of grey-green jade, tinted by the shallow bottom, the branched corals and the variety of their soft, pastel colours, around which swim schools of little blue, yellow, and brown-striped fish of a material comparable to enamel. The ensemble is truffled by the blackish-brown sea cucumbers, almost inert, languid. This lagoon allows the diver, a painter on holiday, to analyse the particular features of the light of the scene, under water and above, in successive impressions by immersing his head and withdrawing it sharply, and to seek the connection between the pale gold of the former and the willow-green of the latter.[15]

Sketched in pencil during that visit, Matisse's split-focus view of the floor of the Fakarava lagoon and the palms and clouds above (illus. 116) mirrored his verbal description and that of Darwin. In his book *Jazz* (1947), he followed his exclamation 'Lagoons, won't you be one of the seven wonders of Paradise for painters?' with the abstract cut-out *Le Lagon* and its colours and contours of the clouds and corals and water. His painted découpages from 1946 and subsequent wool tapestries *Polynesia, the Sea* and *Polynesia, the Sky* included stylized biomorphic shapes of corals, medusae and other tropical marine organisms

Opposite page: 116 Henri Matisse, The Apataki Channel *('Drawing illustrating a letter to the author, which with Matisse's agreement, has been enlarged to occupy two pages in* Poésie 42*'), in [Louis] Aragon,* Henri Matisse, roman, *vol. 1 (1998). Above: 117 Henri Matisse,* Polynésie, The Sea, *1946, paper cut-outs, heightened with gouache and mounted on canvas.*

in expanses of shallow and deep marine blues (illus. 117).

The British marine ecologist T. A. Stephenson, an accomplished artist,[16] was a member of the Great Barrier Reef Expedition of 1928–9 led by C. M. Yonge. Having viewed the reef through the window of a metal-helmeted diving suit (which struck an Australian journalist visiting the scientists as a vision of the armour-suited outlaw Ned Kelly preparing for his final battle with the police), Stephenson was aware of the optical limitations of rendering simultaneously the underwater panorama and the detail of a coral reef. Like Matisse, he would use a split-focus drawing to show at once and seemingly implausibly the underwater reef front and the exposed reef flat as well as distant palm trees.

Most casual viewers see the colours of the water above coral reefs from the deck of a ship or from a beach (illus. 118). These colours, which appeared in the very different artistic styles of Matisse and Stephenson in the Pacific, and more than fifty years earlier by Ernst Haeckel in the Red Sea off El Tor (also known as Tor and Al-Tur) in Sinai (see illus. 184 in Chapter Six), are remarkably congruent, reflecting their painters' scientific and artistic eyes. Philip Henry Gosse, the naturalist and painter, described the scene at the surface and its colours, including

> the white coral beach, the massy foliage of the grove, and the embosomed lake . . . The colour of the lagoon water is often as blue as the ocean . . . yet shades of green and

yellow are intermingled, where patches of sand or coral knolls are near the surface, and the green is a delicate apple shade . . . These garlands of verdure seem to stand on the brims of cups, whose bases root in unfathomable depth.[17]

The 'delicate apple', 'grey-green jade' or 'a copper flame' colour (according to Gosse, Matisse and Stephenson, respectively) of shallow water seen from above a reef was not imparted by the coral but resulted from the wavelength-dependent optical attenuation and scattering of sunlight by clear seawater and its reflection back to the eye from the shallow off-white, coral-sand bottom. Longer wavelengths (red and orange) are most strongly absorbed by seawater and are filtered out first as light penetrates the sea. Yellow and green wavelengths are less strongly attenuated, and blue is absorbed least. This differential also gives the familiar bluish tint to photographs taken a few metres *under* water without a flash.

Thus, in shallow reef water less than 10 m (33 ft) or so in depth, red and orange wavelengths are absorbed, leaving yellow, green and blue to

Above: 118 T[homas] A[lan] Stephenson, 'View of Part of the Landward Side of Yonge Reef, Outer Barrier, from the Deck of a Motor-launch. The Mountain is Lizard Island, and Beyond it the Queensland Mainland is Visible'. Endeavour, *5 (1946). Opposite page: 119 Sheltered platform reefs, Great Barrier Reef. Photograph by Roger Steene in J.E.N. Veron,* Corals of the World, *vol. III (2000), p. viii.*

be reflected upward from the sandy bottom. In the deeper part of this range, the longer passage of light through the optical filter of the water column removes more of the yellow, enriching in blue-green the ambient light that reaches the eye of a viewer above the water (illus. 119). Also, when water molecules absorb blue light, some of it is deflected (scattered) and loses energy, being converted to slightly longer, less energetic blue-green wavelengths that enhance perception of this colour. Less-filtered light reflecting back from a shallower bottom will appear yellower to such a viewer (illus. 120), particularly because the golden brown corals themselves are visible. Seen from aeroplanes or spacecraft, oceanic atolls appear as

turquoise gems set in a sea of lapis lazuli because as light penetrates deep water away from the reef, even the green wavelengths eventually are absorbed and attenuated, leaving only the blue wavelengths to be scattered back to the observer and giving deep water its Prussian blue colour.

Baron Eugen von Ransonnet-Villez, a contemporary of Gosse and a peripatetic diplomat in the Middle East and East Asia, went underwater to observe the effects of light on the appearance of corals and other objects from a small iron diving bell fitted with a glass viewing port (see illus. 122). Ransonnet, who made his observations and sketches at a few metres' depth, published the first documented images of coral reefs viewed underwater. The submarine colour shift is immediately noticeable in his lithographs. He described the underwater palette in a book of 1867,[18] noting the bluish-green hue that prevailed, the rapid disappearance of red and the surprising brightness of green (which, as we have seen in corals, may involve green fluorescent protein).

Ransonnet illustrated his underwater studies of corals with four full-colour lithographs for his book: the prevailing ambient hue is Matisse's grey-green jade. In Ransonnet's earlier coral reef illustration (illus. 121), done while sitting in a boat at El Tor in 1862, the same grey-green cast illuminates the branching, foliose and tabulate corals, and the soft corals, in the scene, as well as the angelfish in the foreground and the elongated shark in the distance. Scrupulously realistic in his

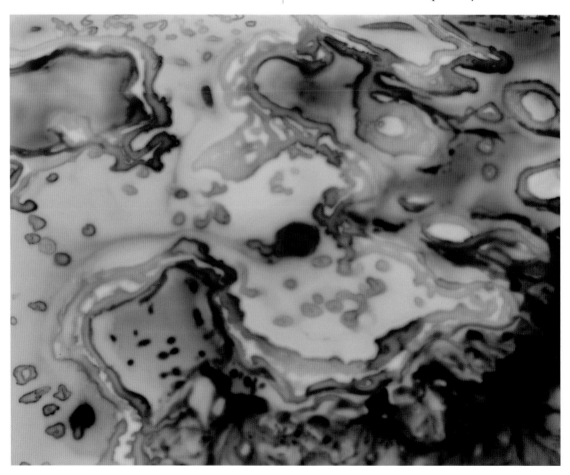

Opposite page: 120 Jaq Chartier, Great Barrier Reef, *2013, acrylic, stains and spray paint on wood panel. Above: 121 Eugen von Ransonnet-Villez,* Coral Reef at Tor near the Port Entrance, *1862, lithograph, retouched by hand by Ransonnet. Right: 122 Eugène de Ransonnet,* Sketches of the Inhabitants, Animal Life and Vegetation in the Lowlands and High Mountains of Ceylon, as Well as of the Submarine Scenery near the Coast, Taken in a Diving Bell, *1867.*

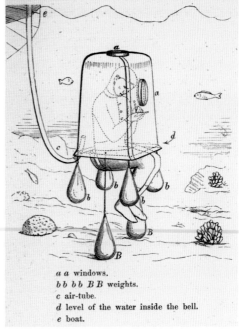

illustrations, Ransonnet noted the greater overall diversity of corals in the wider area and at different depths, but in a given scene showed 'only the species that are really found there'.[19] Haeckel later stylized (as was his wont) much the same scene as painted by Ransonnet, but using characteristically brighter hues without the realistic submarine tint and diversifying the reef life, with the shark moving menacingly nearer (illus. 123).

After Ransonnet, but before Stephenson and Matisse conveyed the colours of corals they saw beneath the sea, that realm had been painted – underwater, again – by Zarh Pritchard in the first quarter of the twentieth century. Early on he weighted himself with a coral rock (reminiscent of Gilgamesh's descent to collect the 'prickly rose'), plummeting to the bottom to make fleeting sketches, and later used a helmeted diving suit for longer working dives in Tahiti (illus. 124). For Pritchard, 'Nowhere does substance appear beyond the middle distance. Material forms vanish into the veils of surrounding color.' He used this indistinctness to realize his impressionistic underseascapes.[20]

The many visitors to Pritchard's exhibitions included the astronomer Percival Lowell (who had envisioned canals on Mars and for whom the 'wonders of the nether [world] tempt me to revisit the homes of my ancestors') and writer

Above: 123 Ernst Haeckel, Arabian Coral Reef near Tur, Sinai*, in* Arabische Korallen *(1876), plate III. Opposite page: 124 Zarh Pritchard's accounts of painting under water captured the public imagination during the first quarter of the twentieth century.*

Jack London ('Your beautiful work has made me homesick for the South Seas. Beautiful it is and full of wonder').[21]

Curious and Concupiscent Corals

Centuries earlier, the dried skeletons of wondrous corals had figured in cabinets of curiosities, the *Wunderkammern*, pictured in period engravings. Several examples suffice. The frontispiece to Ferrante Imperato's *Dell'historia naturale* of 1599 showed various corals, including *Dendrophyllia ramea*, atop shelves in Imperato's *museo*. Levinus Vincent's 'coral cabinet' of a century later, reproduced almost as often as Imperato's

45ᵉ ANNÉE — N° 2317   HEBDOMADAIRE : **20** CENTIMES   DIMANCHE 21 AOUT 1921

Le Pèlerin

REVUE ILLUSTRÉE DE LA SEMAINE

Abonnement Annuel :   Abonnement annuel combiné :
FRANCE... **10 fr.**   ÉTRANGER... **12 fr.**   Croix et Pèlerin............ **48 fr.**

RÉDACTION ET ADMINISTRATION — 5, RUE BAYARD — PARIS - VIIIᵉ

Le peintre Zarh Pritchard est arrivé à peindre au fond de la mer, dont la richesse de décor est incomparable. (Dessin de LECOULTRE.)

frontispiece, was more taxonomically refined, reflecting growing scientific knowledge and specialization (illus. 126).

About 1700, Peter the Great began to establish his *Wunderkammer* of *naturalia* in St Petersburg, expanding it substantially in 1717 by purchasing the Dutch preparator Frederik Ruysch's collection of thousands of curious objects.[22] Ruysch arranged many of his specimens in disturbing tableaux that often included human foetal skeletons posed in miniature landscapes, typically with symbols of death or the transience of life. Rocks in such landscapes might be kidney- or gallstones, and he used his knowledge of medical anatomy and skill in specimen preparation to craft branching 'vegetation' from the ramifications of the human vasculature.

Marine specimens were preserved in liquid, and the jars' stoppers also were elaborately adorned. Ruysch's description of his first cabinet lists black Ambonese lithophytes (antipatharians, which strongly resemble the human vasculature used in his tableaux), branched white (scleractinian) coral and tubular red (organ-pipe) coral. These corals and some vasculature are seen on the stoppers of the jars in plate VI of his *Thesaurus animalium primus* (First Animal Cabinet) of 1710 (illus. 125).

The *Wunderkammern* of princely collectors such as the Electors of Saxony, now housed in the Green Vault of Dresden Castle, became more opulent, including not just objects of *naturalia* such as amber, ivory, *Nautilus* shells, minerals and various corals, but pieces embellished with elaborate gold and silver work. Treasures of the branches of the Habsburg imperial family, too, form some of the most important historical collections in Europe. In the Dresden and the Habsburg troves, works incorporating precious red coral are touchstones of Renaissance and Baroque decorative art, often reflecting coral's symbolic meanings and mythological underpinnings. The artefacts' pop-cultural appeal is evident as they cram social-media-

Above: 125 Plate VI in Frederick Ruysch's Thesaurus Animalium Primus/Het Eerste Cabinet der Dieren (First Animal Cabinet), *1710, showing the stoppers of two jars decorated with specimens of antipatharian, scleractinian and organ-pipe corals, as well as shells, fishes and blood vessels. Opposite page: 126 Eighteenth-century coral cabinet.*

cum-photo-sharing sites such as Pinterest and Facebook and become accessible, albeit not always in context, to audiences far greater than ever envisioned by the collectors.

In 2011 the artist Mark Dion revisited the original awe-inspiring, imagination-piquing and educational intent of cabinets of natural and artificial curiosities and created one 11 m

high by 18 m wide (36 × 59 ft) as an encyclopaedic centrepiece of his *Oceanomania* project at the Musée Océanographique de Monaco. In it he used objects from the collections begun by Prince Albert I, including natural history, fine art, anthropological artefacts, seagoing and undersea technology – 'things alternately curious, pretty, surprising, artistic and scientific'[23] – and not least, an array of corals. Dion's cabinet reiterated the relationship between art and science, the guiding principles on which Albert I had founded his museum as a 'Temple of the Sea', dedicated as such in 1910.

Opposite page: 127 Anne Vallayer-Coster, Sea Fans, Lithophytes and Seashells, *1769, oil on canvas. Below: 128 Domenico Remps,* Still-life in Deception, or Cabinet of Curiosities, *1690s, oil on canvas. Overleaf: Left, 129 Wenzel Jamnitzer,* Daphne, *1570–75, gilt silver, red coral, stone; Right, 130 Daniel Arnoul,* Chevalier, *1990, 18K gold, red coral, sterling silver, amethyst, patinated bronze, ebony inlaid with lapis lazuli.*

Corals and other zoophytes likewise held a fascination for painters. The most painterly assemblage of diverse calcareous fauna, including Mediterranean red coral and related gorgonians, as well as sponges and mollusc shells, remains the masterfully composed and coloured still-life from 1769 *Panaches de mer, lithophytes et coquilles* (Sea Fans, Lithophytes and Seashells) by Anne Vallayer-Coster, a court painter to Marie Antoinette. It is a coral cabinet 'outside the box' (illus. 127).

Earlier, *Natura morta a inganno* (Still-life in Deception), the trompe l'œil painting of assembled *naturalia* and *artificialia* dating to the 1690s and attributed to Domenico Remps, exemplified seventeenth-century cabinets (illus. 128). Whether or not being intentionally systematic, Remps gathered here representatives of three major taxa of corals: the sclerotized black antipatharian and the heavily calcified, white, reef-building scleractinian (both in Hexacorallia), as well as the precious red coral (a gorgonian, in Octocorallia). The branching form of the red and black corals would remind the viewer of their duality as trees

Opposite page: 131 Giuseppe Arcimboldo, Water, *1566, oil on panel. Right: 132 Coral and bells, English, 18th century, silver and red coral.*

from the ocean, even the tree of paradise in the sea written of by Paracelsus.

The symbolic relationship between trees and corals had been explicitly visualized by the Nuremberg goldsmith Wenzel Jamnitzer in his gilded silver statue of the nymph Daphne being transformed into the laurel tree, executed around 1570–75 (illus. 129). In Jamnitzer's hands, the nymph metamorphosed not into a literal laurel, but rather one embodied as red coral. Daniel Arnoul placed atop a triumphant knight another red coral tree having golden leaves of both oak (for strength and longevity) and laurel (representing victory and awards) (illus. 130).

Also prominent in Remps's painting is a human skull, with a striking colony of red coral emanating from its forehead. Joseph Pitton de Tournefort mentioned seeing such an object in the *cabinet* in Pisa in 1700.[24] The image instantly recalls Giuseppe Arcimboldo's painterly potpourri *Water* of 1566, with red coral jutting from the forehead of a human figure composed entirely of marine organisms and the figure itself constituting a corporeal curiosity cabinet (illus. 131).

Arcimboldo was adept in semiotics so there probably is import in the coral branch and its elevated position, painted when he was in the employ of Holy Roman Emperor Maximilian II. The coral might symbolize a lofty association with wisdom and spiritualization of inner power and sovereignty, akin to the 'third eye' in the forehead in Buddhist and Hindu religion or to the rearing cobra (Uraeus) on the forehead of Egyptian rulers.

There also may be a bawdy interpretation, unintended by Arcimboldo, which nevertheless would have struck the contemporary viewer of the conspicuous coral. Its prominent position, convoluted form and redness are clearly reminiscent of a cockscomb (a simile that also occurred in classical Chinese literature), which in the European Middle Ages had an explicit phallic connotation. Not only had coral been worn as an amulet, as well as a love charm associated with fertility, but 'the material was shaped into phallic charms (*fascina*) to be hung about children's necks [and used as teething aids], sometimes equipped with bells . . . The bells allow [one] to quibble on testicles'[25] (illus. 132).

Henry Wadsworth Longfellow wrote of the innocent use of the 'coral and bells' in his poem 'To a Child' (1845), mindful also of the red coral's marine origin (albeit in the wrong sea):

With what a look of proud command
Thou shakest in thy little hand
The coral rattle with its silver bells,
Making a merry tune!
Thousands of years in Indian seas
That coral grew, by slow degrees,
Until some deadly and wild monsoon
Dashed it on Coromandel's sand!

But less innocently, in 1791 the coral and bells had figured in a cartoon satirizing the marriage of a British baron to the widow of the banker Robert Child (illus. 133). Here, the new husband dandles the widow on his knee while she clutches the coral, whose bells are replaced by bulging moneybags. In the cartoon, Baron Francis Ducie is called L[ord] Ju[i]cy; he was a captain in the Royal Navy after whom an island in the Pitcairn group was named, the refuge of the mutineers from HMS *Bounty* and now a large marine reserve.

Left: 133 L– Jucy Dandling His Angelic Child, *1791, hand-coloured etching (artist unknown).*
Right: 134 Pablo Picasso, Humorous Composition: Jaume Sabartés and Gita Hall May, Cannes, 31 July 1957, *coloured grease pencils on magazine printed paper. Opposite page: 135 Odilon Redon,* Blood Flower, *1895, pastel on grey paper.*

The child's teething devices apparently were put to other, sexual and metaphorical, uses.[26] In 'A Pleasant Jigg between Jack and His Mistress', a late seventeenth-century ballad collected by Samuel Pepys, a woman tries to seduce her husband's coach driver, telling him, 'Thy Corral and Bells, and Whistle . . . For pleasure exceeds the [b]est in the Town.' François Rabelais in his sixteenth-century tales *Gargantua* and *Pantagruel* clearly intended 'a fine eleven-inch-long branch of red coral' being brought home by a traveller for his wife's Christmas box to mean a phallus, particularly because elsewhere one of Gargantua's nursemaids refers to his penis as her 'coral'. The simile was also common in England at the time.

Less overtly prurient than in Rabelais, but not always innocent, polished red coral mirrored human anatomy as the lips of poets' lovers. In sixth-century China, Xu Ling described a coquettish woman in 'Wanzhuan-ge': 'How beautiful she is, the way she hides her coral-like lips behind her fan.'[27]

The simile recurred for centuries in European poems, including those of Cervantes in his pastoral novel *La Galatea* (1585) and Shakespeare in 'Sonnet 130' (1609). The conceit became hackneyed enough for Umberto Eco to have Father Caspar disparage it in *The Island of the Day Before* (1994):

What were these corals . . . which he [Roberto] knew only as jewels that according to poets had the color of a beautiful woman's lips?

Regarding corals Father Caspar remained speechless . . . The corals of which Roberto spoke were dead corals, like the virtue of the courtesans for whom libertines employed that trite simile.

In Shakespeare's *Venus and Adonis* (c. 1590), it was the young male lover who had the moist, 'sweet coral mouth'. James Gates Percival, in 'Retrospection' (1823), combined the image of lips and blossoms: 'And her coral lips part, like the opening of flowers.' Coral's sensuality would often be repeated (illus. 134).

Odilon Redon, much influenced by the undersea life he viewed in the aquarium of

France's earliest marine station at Arcachon, often explicitly incorporated marine organisms into his Symbolist paintings beginning about 1900. Although we do not see an actual coral in his pastel *Fleur de sang* (Blood Flower; 1895), the work conjoins the imagery of red coral (represented by a literal red flower), a watery setting and the soft warmth of female skin (illus. 135), almost exemplifying Jules Michelet's exchange from his popular book *La Mer* (The Sea; 1861):

> 'Madame, why is it that you prefer this tree of a dubious red, to all the precious stones?'
> 'Monsieur, it suits my complexion. Rubies are too vivid, they make me look pale, which this, somewhat duller, rather more favorably contrasts fairness . . . The coral has something of the softness and even the warmth of the skin. As soon as I put it on, it seems to become part of myself . . . I know . . . that its Eastern and true name is "the Blood-Flower".'[28]

A decade later, Émile Zola set a living tableau in a treasure cave where each lady wore jewellery of the material her character represented. Here, 'Countess Wanska lent her somber passion to Coral even as she lay with upraised arms weighed down by red pendants like some monstrous but fascinating polyp exhibiting a woman's flesh in the gap between nacreous pink shells.'[29] For many, the colours of coral indeed 'were like the pinks and reds of life itself, of the living flesh'.[30]

Two years after Michelet's book was translated into English, Dante Gabriel Rossetti painted *Monna Vanna* (Vain Woman, 1866), which more literally recalls Michelet's exchange than does Redon's allusive pastel. Rossetti frequently used red coral jewellery in his paintings of sensual, often fair-skinned and auburn-haired, women (illus. 136).

Such femmes fatales would become macabre in late nineteenth-century Symbolist art. The visage of Medusa, 'a screaming head with bulging eyes . . . is the manifestation of an art that forces the viewer to confront his obsessions'.[31] Examples of this visage include Lucien Lévy-Dhurmer's pastel *Méduse* or *Vague furieuse* (Medusa, or Furious Wave) of 1897, in which the Gorgon, screaming in a breaking wave, rends her breast with a clawed hand, while some of the seaweeds that are her hair have turned red (illus. 137).

Coral Riffs

Its manifold allusive, evocative and symbolic richness saw coral capture the imagination beyond literature and the visual arts. Apart from his rapturous visual descriptions of corals and the reef, Joseph Beete Jukes described in the first volume of his *Narrative of the Surveying Voyage of HMS Fly* (1847) being moved by a sonorous experience there:

> The unbroken roar of the surf with its regular pulsation of thunder, as each succeeding swell first fell on the outer edge of the reef, was almost deafening . . . Both the sound and the sight were such as to impress the mind of the spectator with the consciousness of standing in the presence of an overwhelming majesty and power.

Melville, too, in *Omoo*, alluded to the musical majesty of the surf on the reef: 'As we went on, the reef-belt still accompanied us . . . thundering its distant bass upon the ear.' Matisse's time in a Tahitian lagoon was an epiphany that forever coloured his art, but he went beyond the visual, to 'the great emotion of solitude that dominates this ensemble, comparable to that caused by the great nave of a Gothic cathedral (Amiens) where the rumbling of the lagoon would be replaced by that of the organ'.[32] Jules Verne also used the

Opposite page: 136 Dante Gabriel Rossetti, Monna Vanna, *1866, oil on canvas.*

imagery of organ pipes in *Twenty Thousand Leagues under the Sea*, where *Tubipora musica* in the Red Sea was 'just waiting for a puff from the god Pan'.

The first music patently referencing coral may have been the neo-medieval two-part canon composed by the seventeenth-century Rosicrucian Michael Maier. He used it embellish the emblem of the alchemist's red coral in his *Atalanta fugiens* by reiterating the Ovidian version of coral's origin. As such, the music served a different purpose from later coral compositions, which sought to establish an evocative mood or merely to play on coral's cultural popularity.

As did the aquarium, the piano became a fixture in Victorian parlours as a popular home entertainment. Victorian-era and early twentieth-century vocal and piano music included the then popular passions for red coral jewellery and idyllic tropical coral islands, titled explicitly and sometimes but not always composed evocatively. Thus Charles Milsom Jr's *The Coral Waltzes* (1849) and Septima Moulton's *The Coral Mazurka* (1896) were lively, listenable compositions but struck no obvious coral chords, the titles apparently playing on coral's being in vogue (illus. 138).

Alphonse Leduc's *Le Collier de corail* (The Coral Necklace, 1861), a schottische, was composed near the height of the red coral jewellery and pianoforte eras. It has a dance hall feel and tempo, but no clear coral connotations, although its rondo form may bring to mind a necklace.

Bryceson Treharne and Zoë Atkins touched many of the familiar bases in *Corals: A Sea Idyll* (1919), for piano and solo voice. The piece was to be performed 'in a gently flowing style', and the singer's coral beads came from a South Sea cave, a 'strand of rose' given by a mermaid who also had 'sea flowers' in her underwater castle (see illus. 6).

Robert M. Ballantyne's enduringly popular novel *The Coral Island* (1858) probably inspired Arthur W. Pollitt's composition *Coral Island* (1922),

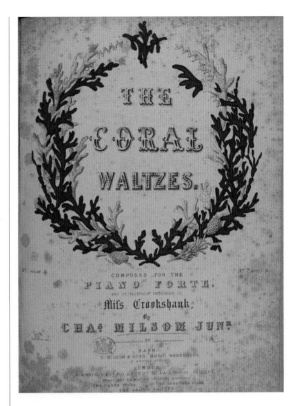

Opposite page: 137 Lucien Lévy-Dhurmer, Medusa, or The Furious Wave, *1897, pastel and charcoal on paper pasted on cardboard. Above: 138 The vogue for precious coral in jewellery and literature in the mid-19th century also extended to popular music.*

a suite of six evocative musical sketches, each having a counterpart in the novel. For example, 'The Wreck on the Reef' alternates between the musical manifestation of the majesty of a sailing ship on the high seas and rolling breakers on the reef front; and 'The Solitary Palm Tree' tosses its tresses in the way that would captivate Matisse.

In Edward Elgar's 1899 orchestration of Richard Garnett's poem 'Where Corals Lie' (1859, see Chapter Three) as part of Elgar's song suite *Sea Pictures*, the possibly suicidal poet drowned, drawn by the alluring beauty of 'the land where corals lie'. The delicate piece ranges from the sunlit shallows amid soaring strings to the darker depths, where the refrain occurs in the lowering registers of strings (some plucked), woodwinds and the mezzo-soprano's voice.

Horace Keats (1895–1945), inspired by the land- and seascapes of his adopted Australia, rendered poems describing these scenes as soundscapes in setting them for piano. In John Wheeler's sonnet 'The Coral Reef' (set to music in 1940), grey shapes slide through channels where 'a dim green light pervades', revealing pillars and colonnades in an architectural guise long fashionable, as we will see, in visions of coral reefs. Elegantly sparse harmonies evoke 'pink white temples' and the 'fragile towers' of living coral.[33] Keats went far beyond a mere accompaniment for Wheeler's sonnet in an ambitious musical composition of ten pages that interwove the piano with the vocal line and conjured the shifting currents and dancing light of reefs (illus. 139).

The Japanese composer Toru Takemitsu (most familiar for his film scores) combined traditional Japanese and Western music and instruments. Takemitsu's works often are aural kaleidoscopes of timbres and silences, exemplified in his early composition *Coral Island (An Atoll)* of 1962, for soprano and orchestra. The album's liner notes tell us that the sonorous opening orchestral part conjures 'the play of the surf ringing an atoll'. The soprano sings poems by Makoto Okha, in the first instance intersecting with the instrumental line as 'a magic sound-space of colors', perhaps envisioning the bright and noisy shallow reef front beneath the surf. The second poem recalls a crystalline coral reef as a bright grove,

> Where sun penetrates into my wood of shells
> I become a transparent coral island . . .[34]

On the jacket of the 1969 RCA Victor recording, the stylized image of pink and black corals by James Alexander recalls the fluid, coral-like patterns in the decalcomania prints by Óscar Domínguez and Max Ernst.

Multifaceted corals lend themselves to the diverse treatments of jazz, although the images summoned by the word 'coral' are not always evident in the music. The trumpeter Neal Hefti (best known for composing the themes for the popular 1960s television shows *Batman* and *The Odd Couple*) included his composition 'Coral Reef' (1953) as the opening track on the album *Swingin' on a Coral Reef* (on the Coral Records label). The piece seems scarcely to riff on reefs, apart from an appealing bouncy swing with a tinkling piano throughout that may for some reflect the teeming life and activity of a coral reef, but might be heard by others as bustling Los Angeles traffic heading to the beach on a sunny day.

Changes in their growth trajectory and colony morphology in response to varying illumination and water currents are a sort of improvisation by corals, and Keith Jarrett, the greatest improviser in jazz piano, may have picked up on this when he composed the ballad 'Coral' (1973) for the vibraphonist Gary Burton, another notable in jazz improvisation. Later, a reviewer wrote of John Scofield's 2000 guitar rendition that his performance 'meanders through "Coral"', which was just what the delicate music, like a leisurely appreciation of a reef by a diver, called for. On the Burton Quartet's reunion CD, *Quartet Live!* (2009), Burton and the guitarist Pat Metheney shared the melody.

Towards the end of the heyday of jazz fusion, Bill Evans (the saxophonist, not the pianist) led a crossover group and included his composition 'Reef Carnival' on the album *Living in the Crest of a Wave* (1984). Evans, playing synthesizer and alto saxophone, gave the piece an electric, almost neon feel that conjured the image of fluorescent corals seen in René Catala's colour photographs in *Carnival under the Sea* and the more recent videos by Miami's science–art cooperative Coral

139 Extract from the score of 'The Coral Reef', composed by Horace Keats (1895–1945), with text by John Wheeler (1901–1984).

Morphologic. The at times Latin rhythm and melody of the *carnaval* produced a feel that is tropical, but not necessarily like a reef.

The Puerto Rican tenor saxophonist David Sánchez combined jazz, Latin and classical music in his album *Coral* (2004), titled for the second movement, 'Coral (Canto do Sertão)' (Coral: Song of the Wilderness), of the Brazilian Heitor Villa-Lobos's composition *Bachianas brasileiras #4*. Coral can be whatever the composer and arranger want it to be, and this music is moody and rich, and nearly as evocative of a reef as Elgar's, even without Garnett's explicit poem, in part because Sánchez augmented his sextet's playing with a symphony orchestra.

The Chilean poet Pablo Neruda was drawn to the sea and knew its denizens well, having a large collection of seashells and other vestiges, and sometimes described himself as 'shipwrecked'. The composer Osvaldo Golijov used Neruda's transformative poetry as the text for his major work for orchestra, double chorus and boys' choir, *Oceana* (1996, revd 2004). The seventh and final movement, 'Coral del Arrecife' (Chorale of the Reef), puns on the Spanish word *coral*, with its ambiguous meaning of both coral and chorale in English. Corals themselves did not appear in the poem's text, but rather it was 'the shells of the reef', including *Spondylus* and *Murex*, both regally tinted like the noble red coral. Of the double chorus, the programme annotator for the Atlanta Symphony Orchestra says, 'the voices hypnotically recall ancient images of reefs and shells and seafarers.'

The sea and its creatures – humpback whale, manatee, seahorse, nautilus, dolphin, manta ray – as well as coral have attracted composers of New Age music seeking to reassure, relax and soothe listeners with atmospheric and haunting moods. Stuart Mitchell thus employed the celesta, harp, clarinet, flute and more in his 'Coral Fugue' in search of a musically mystical marine biology with a modern molecular mark: in *DNA Variations I – The Sea* (2010), the fugue is a 'DNA interpretation of [a] coral gorgonian sea fan'.[35] Edward Weiss composed 'Coral Reef: Take an Underwater Journey!' (2008) specially for newagepianolessons.com, and it is decidedly watery, emphasized by lots of pedal.

Kate Bush is no stranger to submarine imagery. In an early composition, coral covered the blue city of 'Atlantis'. Later, in 2005, her haunting ballad revisited 'A Coral Room' in a ruined undersea city, draped in a fisherman's net, amid sunken masts that formed a seafloor forest. The intensely personal song dealt with the passage of time and things no longer there, and the death of her mother, set in the land where corals lie.

5 Coral as Commodity

> Fishing for coral is entirely special; it has nothing analogous with any other fishery in our European seas. This is due to the very nature of the product it supplies.
> Henri de Lacaze-Duthiers, *Histoire naturelle du corail* (1864)

> The first man to locate these [caverns] . . . is certain to have his efforts repaid by the discovery of a whole new forest of coral. Then the same happens as with forests on land, for after long labour with the necessary apparatus in these caverns, when everything has been cut from them, several decades must elapse before the forest is renewed.
> Luigi Ferdinando Marsigli, *Histoire physique de la mer* (1725)

In the epigraph above, Henri de Lacaze-Duthiers neglected to mention one commonality among different fisheries: the eventual tendency towards overexploiting the resource. Still, Lacaze-Duthiers would propose far-sighted management and controls of coral fishing similar to those used in fin fisheries, and Count Marsigli had recognized early the necessity of a rest period for red coral populations to recover from being cut. By the time of these writings, precious red Mediterranean coral had been favoured for millennia to embellish special objects and as personal adornment and jewellery in the form of beads and rosaries, brooches, cameos, amulets and tiaras, as well as nostrums, gifts to Buddhist Arhats in Tibetan monasteries and luxe *objets d'art*. When Pliny the Elder wrote of it in the first century of the Common Era, Rome was already having trouble meeting the Indian demand for precious coral. As non-selective fishing gear became more efficient (and devastating), the coral industry in the Mediterranean went through local cycles of boom and bust that eventually were repeated in precious coral species in the Pacific, fished primarily by Japan, Taiwan and China. Their place as luxury goods that command a high price and renew slowly has driven the worldwide history of precious corals as commodities gathered and fashioned by artisanal, often family businesses.

Coral BCE

The scattered, unreferenced mentions that the cultural occurrence of red coral dates as far back as 25,000–30,000 BCE seem an exaggeration. References to such ancient, Palaeolithic (Old Stone Age) usages of red coral are conspicuously absent from modern scholarly works, and in the case of the artefact 'supposedly more than 25,000 years old', 'there is no guarantee . . . that this relic is in fact precious coral.'[1] Other archaeologists also noted the uncertainty of such very early materials.[2]

Burial in acidic soil, or burning, can change the colour of red coral, making it harder to recognize, so that it may be overlooked.[3] Archaeological sites on the Palaeolithic coastline of the Mediterranean that might contain red coral would now be underwater following post-glacial sea level rise. For example, the opening to the Cosquer grotto near Marseille, with paintings dating to about 25,000 BCE, is now 37 m (120 ft) beneath the surface of the sea. Still, there is no hard evidence for the Palaeolithic use of red coral in western Europe.

The earliest documented occurrence of worked beads of precious (pink or amber-

FIG. 312. POTTERY WITH MARINE DECORATION, L. M. I b. a, 'MARSEILLES' EWER, CRETAN, FOUND IN EGYPT; b, KNOSSOS; c, GOURNIÀ; d, e, PALAIKASTRO; f, PSEIRA.

coloured) coral may be from graves in the Lower Neolithic (New Stone Age) city of Çatalhöyük in Anatolia (western Turkey, bordered by the Mediterranean, Aegean and Black seas), which dates to the seventh millennium BCE.[4] Other examples of Mediterranean *Corallium rubrum* (and its rare white variant) fashioned as jewellery have been found at sites in Italy, Switzerland and Germany dating from between 5265 and 3440 BCE, evidence of long-distance European trade during the Neolithic.[5]

At Hallstatt Iron Age sites (eighth to sixth centuries BCE) in the Danube, Rhine and Marne valleys, Mediterranean coral appeared as beads and pendants for personal adornment and as an accent to metal objects such as fibulas (brooches or clasps for cloaks).[6] Later, in the La Tène culture (fifth century BCE), the use of coral by the Celts peaked with their embellishment of iron and bronze objects including fibulas, belt buckles, torques and flagons as well as weapons and fittings

Left: 140 The pieces of precious coral inlaid in this La Tène bronze flagon of c. 450–400 BCE have turned beige and white during their long burial. Right: 141 Minoan pottery decorated with dolphins, octopuses, sea urchins, gastropods, algae and corals; Fig. 312 in Sir Arthur Evans, The Palace of Minos at Knossos, *vol. II, part 2:* Town Houses in Knossos of the New Era and Restored West Palace Section *(1928). Opposite page: 142 Mathurin Méheut,* Service, La Mer, *c. 1920. Presentation board: gouache and graphite on paper. The artist's porcelain dinner service featured red corals and gorgonians among other marine animals.*

on chariots, in the last instance most famously those used in burials of nobles (illus. 140). The lack of coral found north of the Alps at La Tène sites dating from after about 300 BCE may reflect its redirection, owing to more profitable trade, to India, which according to Pliny was absorbing all the available coral by the first century CE.

Red coral in its life position – growing upside down in shallow-water caves and beneath ledges – was used as a decorative marine motif, together with the octopus, argonaut (paper nautilus) and other marine animals and seaweeds, on art pottery of the Minoan culture of Crete (1700–1400 BCE) (illus. 141). There seem to be no reports of coral fishing, working or trading by the Minoans,[7] possibly because it is less common around Crete than in the western Mediterranean. However, in *Il corallo nella storia e nell'arte* (Coral in History and Art, 1965), Giovanni Tescione noted that examples of these graceful ceramics were found in Egypt and Marseille, and indeed pottery was among Minoan exports. In the hands of Minoan potters, even the stylish image of red coral, if not coral itself, was a valuable trade commodity.

Red corals (and gorgonians) resurfaced in early twentieth-century France as motifs on the faience ceramic dinner service by the decorative artist Mathurin Méheut (who studied marine biology and used many images of marine life in his work). Méheut and the manufacturer, Henriot-Quimper, received the grand prize at the 1925 Exposition Internationale des Arts Décoratifs et Industriels Modernes for the 104-piece set (illus. 142).[8]

The prophet Ezekiel (*fl. c.* 600 BCE) enumerated the goods traded by Phoenicians at Tyre, in Lebanon. Although they traded for red coral elsewhere in the Mediterranean and left caches of valuable carnelian, agate, ivory and red coral at Carthage in North Africa, according to the Book of Ezekiel (27:16) they also received another coral from Edom (called Idumea by the Romans) in the southern Levant. Edom's only port was near today's Eilat or Aqaba, at the northeastern tip of the Red Sea, which does not harbour *Corallium rubrum*. Most translations of Ezekiel mention simply 'coral', but at least one (the New English Bible) specifies 'black coral'. This makes sense,

for although the Red Sea lacks red coral, it does have antipatharians. Chinese records likewise report ancient trade in black coral from the Middle East. Both red and black corals appear in Tibetan *thangkas* (narrative paintings intended as teaching tools or aids to contemplation) (illus. 143).

Mediterranean coral did not occur in ancient Egypt until it was brought there by the Greeks in the seventh century BCE,[9] and did not see significant use until Ptolemaic times (332–30 BCE).[10] On the golden throne of Tutankhamun (r. 1332–1323 BCE), the exposed skin of the figures of the young pharaoh and his wife sometimes has been taken for red coral (for example, in Giovanni Tescione's treatise *Il corallo nelle arti figurative,*). However, according to Zahi Hamwass, former Egyptian Minister of State for Antiquities Affairs, in his book *King Tutankhamun: The Treasures of the Tomb* (2008): 'The exposed skin of both king and queen has been inset with dark red glass.' There is no mention of red coral in Hamwass's book; the red in some of the spectacular pendants and other personal items of the pharaoh included coloured glass, carnelian and unspecified 'semiprecious stones'. Nor did Cyril Aldred's standard work *Jewels of the Pharaohs* (1971) refer to coral in the making of Dynastic Period jewellery. Considering the mystiques that have developed separately around red coral and Dynastic Egyptian civilization, the two inevitably have become associated, but the physical evidence seems scant.[11] Before the earliest Egyptian use of precious coral, simple beads fashioned by breaking apart red, tubular organ-pipe coral (*Tubipora musica*) from the Red Sea appeared in the prehistoric Badarian culture in Egypt as early as 4400 BCE.

The Silk Road Paved in Red Coral

Pliny claimed that Roman exportation of Mediterranean red coral to the East was already taxing local supplies by the first century CE, and ultimately Rome's trade with India was a drain on her resources. Zoroaster (*fl. c.* 1000 BCE), the humanistic religious reformer who lived on the eastern Iranian Plateau, wrote of the magical protection against evil afforded by apotropaic red coral, as well as its beneficial effects against many ills. Thus, red coral evidently had reached Central Asia by that time. Coral was already known in China by the second century BCE, when the word coral (*shanhu*) appeared in a poem.[12] By 81 BCE, red coral had crossed the frontier in caravans and was recorded as a national treasure in the court of the Western Han dynasty.[13]

The anonymous author of the *Periplus of the Erythraean Sea*, writing at about the same time as Pliny, sailed regularly between the Red Sea and India along part of the maritime Silk Road in the Arabian/Persian Gulf and the larger Indian Ocean.[14] He listed Mediterranean coral among the goods traded. Such materials were moved between the Mediterranean and Red seas by overland routes from Alexandria. Some Mediterranean coral was exported via Cane (Qana) on the southern coast of Arabia, allowing a long-lived jewellery industry to develop in Yemen (illus. 144).

Roman and Greek merchants sailing out of the Red Sea traded coral for Indian pearls and Chinese silk thread at Barbarikon (near the mouth of the Indus River in present-day Pakistan), and further southeast along the Indian coast at Barygaza (Bharuch, north of present-day Mumbai), where coral and other goods were exchanged for black pepper, much valued in Rome. From both ports, coral was carried inland, eventually joining the Silk Road of Central Asia. During the first and second centuries CE in India, a worldly Mahayana Buddhism, displaying an understanding of material wealth in its dealings with foreigners, introduced bodhisattvas –

Opposite page: 143 Arhat Ajita*, central Tibet, late 15th–early 16th century, pigments on cloth. Detail showing elephants uprooting red and black corals.*

worthy humans meriting the status of nirvana but who opted to stay on earth and help others cross the ocean of pain on the *mahayana* (great vehicle) to nirvana. Mahayana sutras specified the gifts appropriate for the Buddha and for invoking the support of bodhisattvas. Among the gifts, which accumulated in Buddhist monasteries, were the Seven Treasures, including coral and semiprecious and precious stones, and other goods mentioned in the *Periplus* that were carried along the Silk Road (illus. 145).[15]

Red coral also was long used medicinally in India and, on its Coromandel Coast, for personal adornment as strings of beads and on sword handles. It also figured in cremation rites, because, according to a seventeenth-century source, 'to leave their dead without ornament of coral is to give them over to the hands of mighty enemies.'[16]

Above: 144 Maarten de Wolf, Yemeni Dress 15, *2014, Sana'a style bridal dress accented with red coral, and a veil of coral beads, which protect against the evil eye.*

Mahayana Buddhism had reached Tibet by the eighth century CE and coral offerings adorned monasteries there. The wider use of red coral in Tibetan culture is evinced by its recommendation as a medical material in the *Four Tantras*, Tibetan teachings perhaps originating in the eighth century CE but more fully elaborated in the twelfth century and incorporating Indian, Chinese and Graeco-Arabic medical knowledge. In the shamanistic Tibetan Buddhism, both coral's red colour and its oceanic origin raised it, as the Tree of Life, to a central position in both religious and everyday objects. In the thirteenth century, Marco Polo arrived at the court of the Mongol ruler Kublai Khan along the Silk Road, having found red coral common enough in Tibet to be used as a medium of exchange.

A problem in interpreting trade records is that 'coral' often signified only a marine product having a treelike form. Song dynasty records say that coral from the Coromandel Coast of India was presented to Tang dynasty officials in 971 BCE,[17] suggesting this was other than precious red coral (which does not grow there), unless red coral first had been traded to India from elsewhere. There has been scant study of the trade in coloured corals indigenous to the Indo-Pacific.[18] By the fourteenth century, Chinese documents listed both the Mediterranean (the Maghreb in North Africa, and Rome) and Calicut on the Malabar Coast of India as 'producers' of coral, but by this time Calicut was a major trading centre, so red coral imported from elsewhere might have been meant. The early fifteenth-century expeditionary fleets of the Chinese admiral Zheng He reached India, East Africa, the Red Sea and the Arabian/Persian Gulf. First-hand accounts reported the trade of coral trees (most likely black coral) in Aden in Yemen, the sale of (red) coral beads and branches in Hormuz on the Persian Gulf (which also had coral trees), the 'production' (probably importation and working) of coral by Mecca, and the trade in coral and production of coral beads in Calicut and elsewhere on the Malabar Coast.[19]

As the English East India Company's red coral made its way by sea from the Mediterranean to England and thence to India and, via the Chinese port of Guangzhou, to the imperial court of the Qing dynasty (late seventeenth to early twentieth centuries) in Beijing, its connotation shifted 'from crop, to commodity, to curiosity, to national treasure'.[20] There, the auspicious red Mediterranean coral, metamorphosed from *corallo*

Right: 145 Detail, Tara Protecting from the Eight Fears, *Kham Province, Eastern Tibet, 19th century, pigments on cloth. Detail shows a monk praying to the goddess Tara for protection from the hazards of water. Behind him are the Eight Auspicious Symbols (conch shell, treasure vase, endless knot, parasol, pair of golden fishes, lotus, wheel and victory banner) and the Seven Precious Gems (including the rhinoceros horn, elephant tusks, orb-shaped gems, crossed gems and a branch of precious coral).*

to *shanhu*, eventually featured in court etiquette and official hierarchy. Officers of the first and second rank wore, in addition to a string of coral and other beads, a red ball or 'button' atop their black velvet caps: in the first grade the button was ruby, and in the second, coral.

Trade in Mediterranean coral extended not only eastward to Asia but also southward to Africa. The spread of Islam around the Mediterranean between the sixth and ninth centuries opened new markets for strung beads of red coral, which saw use by the faithful in enumerating the 99 names of Allah. Early on, the superior coral from Marsa'el Kherez (Arabic for 'port of beads') on the eastern shore of the Maghreb was favoured. By the early twelfth century the geographer al-Idrisi could write of coral of unsurpassed beauty collected in the Maghreb, at his home town of Ceuta on the Strait of Gibraltar, where:

> There is a bazaar occupied with the cutting, polishing, rounding, drilling, and finally stringing the coral. This is one of the main export items; most of it is transported to cities of Ghana and Sudan where it sees great use.[21]

Coral as Jewellery and Ornament

In Europe, 'The adoption of the coral-bead rosary, of highly variable forms and dimensions, led to secular personal ornaments and jewels.'[22] Coral-bead necklaces adorned peoples around the shores of the Mediterranean and beyond. Trapani in Sicily was an early centre of such coral working from the twelfth century. Coral carvings such as cameos also became popular from the fifteenth century, being used in pendants and rings and as focal points in bead necklaces. Religious items became more figurative during the Renaissance. These items included images of Christ and saints, and used coral as the cross where Jesus was crucified and the tree where St Sebastian was martyred (illus. 146).

Gold- and silversmiths, especially from Nuremberg and Augsburg in Bavaria, included

Left: 146 Nativity scene, 17th century, Italian red coral and other materials. Opposite page: 147 Natternzungenkredenz (tonguestone holder). Stand: western Germany or Burgundy, c. 1400; mountings: Germany, 15th and first half of 16th century; gilded silver, red coral, 13 fossilized sharks' teeth, rubies, sapphires. The object has been in the treasury of the Teutonic Order since 1526.

Left: 148 Coral-handled spoon. Below: 149 The Habsburg Archduke Ferdinand of Tyrol's ceremonial sword with a coral handle (armoury in the Hofburg Palace in Vienna). Opposite page: 150 Melchior Bayr, Zocha welcome cup, 1667, gilded silver, red coral.

red coral symbolically in their works. Striking among these is the *languier*, *Natternzungenkredenz* or *Natternzungenbaum*, comprising a gilded silver base used as a salt-cellar and supporting an impressive treelike branch of coral having gilded mountings hung with thirteen fossil sharks' teeth (illus. 147). The piece, dating to about 1500, is named after the teeth, thought at the time to be the tongues (French *langues*; German *Zungen*) of poisonous adders (*Nattern*) and having the power to detect and detoxify poison. In use, the *languier* would rest on a side table or credenza from where guests could take a tooth to test their wine, a folkloric tradition originating on Malta, where the teeth were common and were intertwined with the legend of St Paul's shipwreck there and his subsequent survival of a snakebite.

Red coral, too, reputedly counteracted poisons and was used to detect them. Was this use in the mind of the goldsmith who created the Natturzungenkredenz? Might the potent coral branch be a 'recharging station' to restore the power of a shark's tooth that had been used to purify a cup of wine? Also on the sixteenth-century German table were items of coral-handled cutlery sometimes embellished with turquoise,[23] admittedly uncomfortable to use but desirable rarities and perhaps used to test for poison with one's own eating utensils (illus. 148). Likewise, the Zocha welcome cup, a communal drinking vessel used to receive visitors, was a stag whose antlers were represented by red coral on the removable head (illus. 150). Also uncomfortable to use would have been Archduke Ferdinand of Tyrol's coral-handled sword, but this weapon was ceremonial and unsuited for use in combat (illus. 149).

Farther afield, in Tibet red coral was displayed with the most spectacularly coloured stones, particularly amber and turquoise. So integrated is Tibet's form of Buddhism with its cultural history that coral became both a part of religious symbolism – a representation of the life

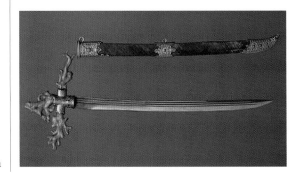

5 CORAL AS COMMODITY

Opposite page: 151 The head of a queen mother, Kingdom of Benin, early 16th century, gun bronze. In life, the mesh around the forward-pointing hairstyle called a 'chicken's beak' would have been made from beads of red coral. Above: 152 Oba Akenzua II greets Queen Elizabeth II on a royal visit. Chief Jeremiah Obafemi Awolowo, first premier of Nigeria's Western Region from 1952 to 1959, stands on the left. On the right is Sir John Rankine, governor of the Western Region from 1954 to 1960. Hand-coloured silver gelatin print by Solomon Osagie Alonge, c. 1956, Benin City, Nigeria.

force and its tie to the far-off sea – and a part of individual attire. Particularly among nomadic groups wealth had to be portable; when worn it also proclaimed the social status of the wearer. Accumulated over generations, the adornments represented a family's heritage of sustained success.

In sub-Saharan Africa, the best-known use of Mediterranean coral has been in the Kingdom of Benin, where its Edo people once held sway over a large part of south-central Nigeria. The beginning of coral's pre-eminence in Benin dates to the fifteenth century, when King (Oba) Ewuare (enthroned in 1440) stole the precious coral from the underwater palace of Olokun, the god of the waters and prosperity. Less mythically, the timing of the appearance of red coral in that oba's accoutrements also coincided with the arrival of the Portuguese by sea, ready to trade

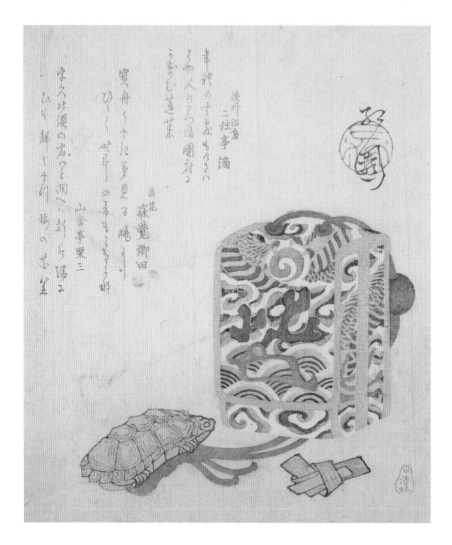

coral, brass bracelets, red flannel and other luxuries, as well as weapons and military protection, for ivory, copper and pepper. Newly wealthy and powerful from maritime trading, Oba Ewuare consolidated the empire that he governed from Benin City.

Among this oba's innovations were his regalia, including crowns, necklaces, shirts, aprons and even shoes made of red coral beads; among an oba's appellations is 'child of the beaded crown, child of the beaded gown' (illus. 152). Wearing these things, the oba became a god,[24] and his regalia of coral beads and red stones were bathed in the blood of a

Above: 153 Nagayama Kōin, Inro, Turtle Netsuke, and Coral Bead*, 1828, polychrome woodblock print on paper. Opposite page: 154 Kuniyoshi Utagawa,* Taira no Tomomori*, 1843–4, polychrome woodblock print on paper. After losing to the Genji in the battle of Dan-no-ura in 1185, leaders of the Taira clan committed suicide, Taira no Tomomori by throwing himself overboard, tied to an anchor. A leader of the Genji became the first shogun.*

human sacrifice (cow's blood is used nowadays) to reinvest them with spiritual power. Ladies of the Benin court also had intricate hairstyles adorned

名高百勇傳 平知盛

平相國清盛の四男にて宗盛の第で智勇
勝れて文武に達し忠孝の道又大兄重盛に
劣らず其死後又是が代りて專ら父清盛を補佐し
從三位に叙し中納言に任じて治承四年四月源三位
頼政友逆の時其討隊の惣大將として指揮号令嚴
密にて誰も恐れ慴き老功の頼政も遂に敗れて宇治にて
自殺し其子仲綱兼網みな討死す其後頼朝義
仲起兵の時知盛種々計を勸むれども宗盛愚
昧にして是を用ひず敗れに敗れて寿永二年七月味方
花洛を退くにも知盛一人が都で死んと怒り
止れども宗盛に従ごふて没落を共より四國
中國西國又は攝州一ノ谷など諸所の新内裡を
築んで姑且先帝を守護せり一ノ谷の勳
功に依れり其子父又陣發せり文治元年二
月義經几波之渡を讃岐八島の棚へ押寄よ
一戰に切勝宗盛父子を容易く生捕し
盛武地に在らむ斯る長州赤間が關の城を攻る
も事中もる也く斯て先帝も既に入水をせて亡のみ
丈八尺許ある大碇を自ら被ぎて海に花入て亡のふ歳四十二と聞ふ

一勇齋　國芳畫

with red coral, as depicted in the 'Benin Bronzes' which are among the most spectacular examples of African art (illus. 151).

Less elaborate than those in Benin, the coiffures of ladies of the Japanese Edo period (seventeenth to nineteenth centuries) were supported and embellished with long hairpins (*kanzashi*) that often featured pinheads cut from Mediterranean coral and with combs that incorporated coral, silver, gold and tortoiseshell. Although much coral was imported into Japan in the early seventeenth century, it was mostly as gifts from the Dutch to the shogun and ranking officials.[25] Coral long remained for the exclusive use of the favoured elite, and 'ordinary people never . . . set eyes on it', but about 1700 coral imported particularly from Holland began to see wider use in *kanzashi* and *netsuke* (illus. 153).[26]

In the nineteenth century, when species of precious corals such as the pink *Corallium elatius* were discovered off Japan, entangled in fishing gear, the industry developed rapidly and Japanese coral, especially in fashionable sculptural carvings, was being exported to Italy by 1887. Domestically, Japanese coral pieces were sold to participants in the Shikoku Pilgrimage to 88 Buddhist temples on that island.[27]

A woodblock print from 1843 shows a large piece of red coral set into what appears to be rock crystal (illus. 154). The branch, and a second piece nearby, serve as a rack for the sword of Taira no

Below: 155 Phillips Brothers (London), Coral Tiara, c. 1860–70, a wreath of sprays and berries on a gilt-metal frame. Opposite page: 156 Mary Lincoln's red coral necklace with cherubs and roses.

Tomomori, a leader of the Taira clan (sometimes called the Heike), defeated by the rival Minamotos (or Genji) in a naval battle in 1185. The ascent of the Genji marked the beginning of feudal Japan, ruled by military leaders, or shoguns, rather than the aristocracy.

Towards the end of the Edo (Tokugawa) era in Japan, Victorian England saw a short-lived vogue for red coral jewellery coincident with the 'aquarium craze' stimulated by natural history writing and growing popular knowledge of marine life. Much of the jewellery, often elaborately carved as flowers, leaves and crosses, or putti and other classical images, came especially from Naples and Genoa, and matching ensembles were favoured by the wealthy. Some coral jewellery was produced domestically: in London, Robert Phillips was famous for carved coral pieces, which included diadems and tiaras (illus. 155). Phillips received the Order of the Crown of Italy for helping to promote the coral industry in Naples. Departing from the Continental imperial style of the early nineteenth century, some of the pieces included coral branches in a naturalistic style derided in 1867 by the bohemian London journalist and pornographer George Augustus Sala as 'twisted sticks of seeming red sealing-wax; and coral earrings had an unpleasant resemblance to . . . the domestic forked radish smeared with red ochre'.[28]

Americans, including Mary Lincoln (Abraham Lincoln's widow, controversial for her expensive wardrobe), also discovered coral jewellery in the mid-nineteenth century. Mrs Lincoln owned a carved coral necklace having two small bouquets of roses, each with a cherub, and a larger, central pendant with a cherub in a wreath of roses (a symbol of everlasting love) hanging below them (illus. 156). Some thought that Mrs Lincoln bought the necklace in the late 1860s 'as a remembrance of her larger angel (Abraham, the centre one), and her two smaller ones (Eddy and Willie) [her deceased sons]'. True story or not, 'That cherub would hang fairly close to her heart.'[29]

One sometimes reads of the use of red coral in ornamentation by pre-Columbian cultures in Mexico. An advance party sent in 1539 by Francisco Vázquez de Coronado in search of the Seven Cities of Cibola encountered Zuni villages in what is now New Mexico, where the people 'wore many coral beads found in the Southern Seas and many turquoise beads they had received from the north'.[30] But red Mediterranean coral would not have reached Mexico until the Spanish brought it. Before that, bright red or orange calcareous decorations were fashioned from the colourful shells of thorny oysters, *Spondylus*, from both coasts of Mexico.

Mediterranean coral, probably first in the form of rosary beads and religious artefacts, made

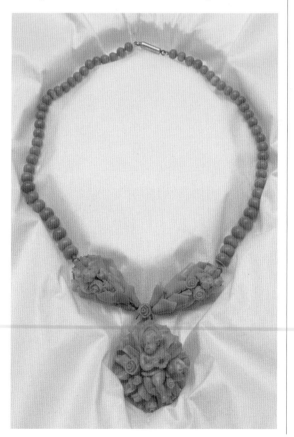

Opposite page: 157 Dan Simplicio, Branch Coral and Carved Turquoise Necklace, *1945, silver beads, branch coral and plaques of Villa Grove turquoise. This Zuni artist was the first to use branches of red coral in their natural form in his distinctive creations. Above: 158 Daniel Arnoul,* The Gorgon's Ring, *1989, red coral branch inset with ivory and entwined with pearled gold.*

its way into the Spanish colonies and missions, and this new red material eventually was adopted for making jewellery, certainly by 1750.[31] By the nineteenth century, Mediterranean coral from Italian companies was traded extensively in the American Southwest, where Hopi snake dancers wore coral necklaces. As elsewhere, coral beads represented a form of wealth, and the length and number of necklaces were a way to display this. Native American craftsmen had learned silversmithing from the Spaniards, and in the early twentieth century the trading post operator C. G. Wallace encouraged them to combine Mediterranean red coral and native turquoise set in silver. Zuni jewellers in particular became adept at producing such pieces, which latterly included coral branches (illus. 157).

If it did not figure much in Dynastic Egyptian jewellery and ornament, precious coral eventually did see use in Egyptian revival works (stimulated by the excavation of the tomb of Tutankhamun in 1922) that were masterpieces of Art Deco. Most recently, contemporary jewellers taking inspiration from marine nature have used coral as both a material and a motif in their opulent designs for wearable sculpture (illus. 158).

Fishing for Coral

The earliest collecting of Mediterranean coral was probably by ancient people picking up fragments that washed up on the strand after storms (see illus. 6).[32] By Pliny's time, Romans were gathering coral by diving and cutting it with a sharp iron tool, as presumably the Greeks and others had done for centuries, and by entangling it in dragged nets. Naked divers working from a steep Sicilian shore and nearby boats appear in a late sixteenth-century print that shows them retrieving large coral colonies, but with no sign of specialized equipment apart from their early version of goggles (illus. 159). Considering the vast quantities exported for centuries to India and beyond, it may seem surprising that coral remained abundant enough to be collected by free-divers, who could work at depths of up to only 20 m (65 ft) or so for one or two minutes. But when Count Marsigli wrote of red coral in his *Histoire physique de la mer*, eventually published in 1725, he said that it could still be found in waters as shallow as 2½ fathoms (4.5 m/15 ft), easily within the range of free-divers.

Inevitably, as the demand for coral grew and many shallow-water populations were thinned, Mediterranean peoples invented mechanisms to extract it from deeper water and on a larger scale as part of organized coral fisheries. Count Marsigli showed two such devices in his *Histoire*

27 Coralium Siculus solers cautusq; specillo   Piscatur: fit demptus aquis durusq; ruberq;
    Ante oculos fixo, placidum cùm stat mare ventis, Ramus; qui tener, et viridis fuit ante colore.

Above: 159 Cornelis Galle I, after Johannes Stradanus, Fishing for Coral, *c. 1596, copper engraving printed on paper. The Latin caption at left mentions the Sicilian divers' using goggles, and at right repeats the view at the time that green coral branches turned red when removed from the water. Left: 160 The* ingegno *or* engin *collecting red Mediterranean coral. Matthys Pool's hand-coloured engraving (fig. 110, detail of plate XXIV in Luigi Ferdinando Marsigli,* Histoire physique de la Mer *(1725)) is from the copy of Marsigli's book that belonged to Sir Joseph Banks.*

– the *engin* and the *salabre*. The *engin* (or Italian *ingegno* – literally, engine or machine) consisted of two wooden beams in the shape of a St Andrew's Cross, weighted and hung with nets to entangle coral branches broken from the substratum by the heavy cross, which was deployed from a boat (illus. 160).

The *ingegno*, mentioned by al-Idrisi in the twelfth century as being deployed from boats crewed by eight men, was adopted, modified and used for centuries throughout the Mediterranean, not only by Arabs in North Africa but also by Catalans, Provençals, Neapolitans and Sicilians in their deep-water coral fisheries. Provençal and Catalan fishers in inshore waters used the *salabre*, with a rasping metal ring and its suspended landing net at one end of a long, weighted wooden beam or iron bar (see illus. 50), to collect coral from its preferred habitat beneath ledges.

In the twentieth century, still larger iron versions of the *ingegno* wreaked havoc on the sea bottom. The particularly destructive *barra italiana* (Italian bar) was a 6-m (20-ft), 1-tonne metal cylinder hung with an array of iron chains each 5 m (16 ft) long that was dragged across the seabed with devastating effect and has been repeatedly denounced by environmental organizations.

The physiological limits to the working depth for free-divers collecting coral, and tantalizing glimpses of the quantities of riches just beyond their reach, made it a risky and dangerous occupation. This limitation led to the adoption of the semi-autonomous *scaphandre*, a helmeted diving suit continuously supplied with air via pumps on the surface.

In the mid-nineteenth century, diving for coral was the principal occupation in the fishing village of Cadaqués on the rocky shore of Cape Creus in Catalonia, not far from Salvador Dalí's birthplace in Figueres (illus. 161 and 162). The utopian socialist and inventor Narcís Monturiol, also a native of Figueres, returning home in 1844 after his youth as a revolutionary in Barcelona, witnessed an accident involving a coral collector in a diving suit who became unconscious while working at 19 fathoms (35 m/ 114 ft) but was hauled

161 and 162 The Greek family Kontos, working from fishing boats off Cape Creus, Catalonia (left), used diving suits to harvest red corals (right). Witnessing a man almost being killed in a diving accident inspired Narcís Monturiol to invent a safer and more efficient means of collecting corals.

Opposite page: 163 and 164 Modern model of Monturiol's Ictíneo II *(top) with the device similar to the Catalan* salabre *mounted forward on the hull for collecting red coral, and (bottom) an image from a promotional brochure showing Ictíneos at work salvaging cannon and collecting coral. Above: 165 Monturiol's submarine association's silk flag (sewn by his wife). The Latin motto 'Plus Intra, Plus Extra' was translated by Robert Hughes as 'Far down! Far out!' in his account of Monturiol in* Barcelona *(1993), and by Matthew Stewart as 'The deeper the better' in* Monturiol's Dream *(2003).*

up and resuscitated. The accident inspired Monturiol to ponder improving the lot of coral divers using a means consistent with his progressive nature and self-education in engineering: a submarine that would extend scientific knowledge of the sea and exploit coral and other undersea resources.[33]

Two versions of his *Ictíneo* (from the Greek *ichthus*, fish, and *naus*, boat; variously New Fish, from *néos*, new)[34] collectively featured innovations such as a double hull (a design still used in modern submarines); a system to eliminate carbon dioxide (CO_2) exhaled by the crew by combining it with calcium and forming a precipitate of calcium carbonate, $CaCO_3$ – a mechanism analogous to that used by corals to deposit their calcareous skeleton; and a steam engine for underwater running that used a chemical reaction to generate both heat for the boiler *and* oxygen gas (O_2), thereby solving the problem of oxygen supply for the crew during prolonged dives (illus. 163 and 164).

Such high technology was expensive, and eventually creditors foreclosed on Monturiol's habitually impecunious submarine association (illus. 165). *Ictíneo II* was sold for scrap in 1868, the year before Verne serialized *Twenty Thousand Leagues under the Sea*, with its fictional high-tech submarine. Despite their successful sea trials and their inventor's best intentions, neither *Ictíneo* ever collected a branch of coral. The use of submarines to exploit corals would have to wait a century.

Use of the diving suit continued well into the twentieth century, but it was the introduction of scuba following the Second World War that brought a new assault on *Corallium rubrum* in the Mediterranean, as told in his personal narrative by Leonardo Fusco,[35] an early adopter of scuba and submarines in coral fishing and, later, conservation. In 1953, Fusco, spearfishing while free-diving, discovered red coral at a depth of about 30 m (98 ft), and returned using borrowed scuba equipment to collect 6 kg (13 lb) of coral. In principle, a scuba-diver gathering coral in relative leisure could collect more selectively than a free-diver during a fleeting, breath-holding plunge to the bottom. Accordingly, Fusco and his scuba-diving colleagues collected thousands of kilograms of coral over the years, especially from Sardinia's then unexploited resource (illus. 166).

Eventually reaching the limit to diving on air, Fusco could see that most of the high-quality coral lay still deeper, between 80 and 120 m (260–400 ft). He became interested in diving using a mixed-gas (helium and oxygen) rebreather, which reduces the decompression problem that comes with nitrogen in compressed air, and 'without the sneaky features of nitrogen narcosis', so by the early 1970s he was making extended heliox dives to depths of 100 m (330 ft), as recounted in his 2001 memoir *Red Gold*.

Revisiting a favourite dive site after twenty years, he found it ravaged and realized that he and

Above: 166 Scuba diving brought a new, efficient assault on precious corals, enabling divers to collect thousands of kilograms of colonies. Opposite page: 167 Painting on the ceiling of the Nishidomari Tenmangu Shrine showing a successful haul aboard a Meiji-era fishing boat under sail, and the trawled nets that entangle coral colonies.

his colleagues had started the process that was completed by other divers. There is no more fervent believer than the convert. Recognizing the devastation being wrought on coral populations, Fusco thought ever more seriously about protecting and conserving them. He turned to submarines, and it was from a mini-sub that he saw at first hand the damage done by the iron *ingegno*, the 'hulking iron cross that plowed across the sea bottom to collide over and over again with the seamount, battering it, and miring it in mud and debris'.[36] Collecting coral by scuba diving had changed, too:

Having discovered every single coral patch there is to find in the sea, the reefs that in the past yielded large quantities of *Corallium rubrum* have now been visited several times by different divers, who strip the last remaining coral in order to avoid coming home empty-handed. Even the ROVs [Remotely Operated Vehicles] . . . with their hydraulic arms . . . can collect the few stumps of coral that escaped the dragnets and the divers.[37]

Despairing at last of the state of affairs in Italy and then Tunisia, Leonardo Fusco got out of the coral business. Late in his career, he used a decompression chamber taken from his last boat to provide hyperbaric medical therapy in a clinic that he founded.

In the Pacific, other precious coral species display various hues: *Corallium secundum* is pink or 'angel skin'; *C. elatius* ranges from cherry red to pink; *Paracorallium japonicum* is oxblood red; and *Corallium konojoi* is white. The first recorded discovery was off Kōchi Prefecture in Japan in the early 1800s, as a by-catch from bottom fishing.[38] Such adventitious coral was bought and sold privately, although one of the feudal domains of the Tokugawa Shogunate restricted such trade and required the reporting of any coral that was fished up.[39] This situation persisted until after the Meiji Restoration of 1868 ended the feudal era. At the end of the Meiji period the coral fishing industry in Japan spread to Kagoshima Prefecture.

Fortuitous entanglement remained the method in the early days when the coral populations were still unexploited. As stocks

became depleted, other beds were located and fished, usually by local people in small boats. Konojo Ebisuya (in whose honour *Corallium konojoi* was named) was the first to promote an organized coral fishery in Japan in the 1830s–40s.[40] He invented more efficient equipment, including a sort of 'mini-*ingegno*' having a bamboo pole fitted with nets, weighted with stones and operated from a boat (illus. 167).

Most of the Pacific precious corals occur below scuba depth, so even in the modern era coral was still dragged up in nets. Fishing eventually depleted more stocks, sending the fishers farther afield, off southern China and Taiwan, where Taiwanese boats also became engaged. In the 1960s, the Japanese and Taiwanese discovered rich banks of *C. secundum* on the Emperor seamounts, northwest of Midway Island, at 400 m (1,300 ft), where they took hundreds of tons of the pink coral. Deeper down, bamboo corals (another family of octocorals) flourish. These semi-precious corals have a more porous, absorbent skeleton, which has led to their being dyed deep red or black as a substitute for, or indeed a counterfeit of, precious red or black coral. Heavy trawling for bamboo coral by Taiwanese, Japanese and Korean boats has devastated the deeper Emperor seamounts.[41]

Coral Cartels
The larger ethnic market for Mediterranean coral provided by Islam led to the prominence of the markets at Ceuta in the western Maghreb and in the east at Marsa'el Kherez (now El Kala and also called La Calle by the French in Algeria). Each had its own local supply of coral as well as that from Sicily, under Muslim control between the late ninth century and its conquest by the Normans in the late eleventh century. Marsa'el Kherez was also a notorious lair for Arab corsairs who plundered the coral catch, as described by François Doumenge:

The resurgence of Christendom . . . in the eleventh century by the recovery of Sicily and the beginnings of the Iberian *Reconquista*, would again modify the coral system. The Christian Mediterranean would establish its control of the sea by the thalassocracies of the large mercantile cities . . . Amalfi in the tenth century, Pisa in the eleventh, Genoa in the twelfth and thirteenth, Barcelona and Marseille in the fourteenth and Naples and Trapani in the fifteenth . . . The basis for their centrality to the coral economy included local products from their own shores, their colonial hold on virgin resources (Corsica, Sardinia, Algarve) and the control of the Maghreb.[42]

In the feudal economy of Catalonia in 1062, the Count of Barcelona reserved for himself a third of the coral fished there. The destruction of Marsa'el Kherez by the Genoese in 1286 opened the Maghreb to coral interests in Genoa, Barcelona and Marseille who negotiated fishing rights that included land bases for their fleets, for example Tabarka Island for powerful Genoese families such as the Doria and Lomellini (illus. 168) and, later, the Bastion of France on the Barbary Coast for Marseille. The coral business typically was a family affair. The exploitation of the Neapolitan coral banks was controlled in the thirteenth and fourteenth centuries by the Anjou family (the Angevins) of France and then by the Aragóns in the late fifteenth century.[43]

Throughout the Middle Ages, coral production was by Christian ship owners while transport was through Muslim lands to meet the demand from India and sub-Saharan Africa, but much of the working, as well as the packing and trade, of coral was in Jewish hands through their hub in Alexandria. In 1492–3 the Jews and forced converts to Christianity (*conversos* or *anusim*) were expelled from the Aragonese territories in Sicily and Catalonia, and from greater Spain when

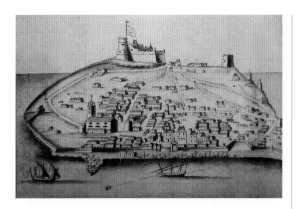

168 An anonymous 17th-century map of Tabarka Island on the coast of Tunisia, showing its Genoese fort and garrison buildings. The island was a concession to the Lomellini family of Genoa (with ties to Andrea Doria) from the Bey of Tunis in 1540. The island was eventually surrendered by the Genoese in 1741, already depopulated because the coral resource was exhausted.

Ferdinand II of Aragón wed Isabella I of Castile. Their exodus, especially from Barcelona, benefited Marseille, Naples and particularly Livorno, which became the hub of the Mediterranean coral trade, replacing Alexandria and Ceuta. Doumenge continued:

> From the late sixteenth to the late nineteenth century, it was Livorno, 'a favourite retirement place for out-of-business English privateers' . . . where a large annual coral fair handled virtually all of the available production.
>
> It is from Livorno and by Livorno that the trading and banking of Jewish communities of Lisbon in the sixteenth century, Amsterdam in the seventeenth, and London in the eighteenth and nineteenth organized their transfer of coral to India by the new sea route via the Cape of Good Hope. The city of London had its Coral Merchants Street, now simply Coral Street.

In India, at Goa, Madras, Bombay, or Calcutta, it was often the parents and relatives of the sender who conducted the sale of the cargo and used the proceeds to purchase precious stones, especially the famous diamonds of Golconda, for Jewish lapidaries of the merchant cities of Western Europe. These networks would be one of the most solid and faithful supports of the East India Company [chartered in 1601] and the British administration of the Indian Empire.

For the three trading centuries, India hoarded, while using huge amounts of coral powder for therapeutic use. It also managed to feed the relentlessly strong Tibetan demand.[44]

Portuguese and Dutch shipments of coral sometimes drove the price down, but the demand was insatiable.

Two Marseille merchants headed the Compagnie Marseillaise du Corail, dating to about 1560, with a network that eventually extended throughout the western Mediterranean, including North Africa (where Peyssonnel would sail with coral fishers in 1725). The Compagnie Royale d'Afrique, controlled from Marseille, took over in 1741 and used La Calle as its base, where the supply of coral, from collection to shipment, fell strictly to company men.[45] Although it monopolized the Tunisian coral trade, eventually the Compagnie was another victim of the French Revolution.

The serially hapless Bourbon monarch Ferdinand IV, III and I (king, sequentially, of Naples, Sicily and the Two Sicilies) granted Paul Bartholomew Martin from Marseille permission to form the Real Fábrica di Corallo in Torre del Greco in 1805, with ten-year exclusive rights reiterated by Joseph Bonaparte. 'It was the decisive step towards the long-sought complete economic cycle for coral . . . [and] . . . led to Torre del Greco becoming the major worldwide producer,'[46] with a shift from trade beads (as in Livorno) to a more artistic tradition.

Fishing by Italians in French-dominated Algeria had become a problem. According to an Italian account, the French were manning their boats with Italians because the French did not have the skills for coral fishing. An English report in the *Practical Magazine* in 1876 made it seem that the crews consisted largely of Italian draft-dodgers in the decades of armed conflicts preceding the unification of Italy in 1871. For Henri de Lacaze-Duthiers, it was a matter of economics, as he documented in the second half of his *Histoire naturelle du corail*. He knew that a sailor in Italy was paid far less than one in France, was less demanding, not self-indulgent when it came to food and as hard-working as the French.

In fact, the work of a *corailleur* (coral fisher) was notoriously gruelling, and Lacaze-Duthiers repeated the common aphorism 'You have to have stolen or killed to become a coral fisher.' Much of the labour involved manually turning the capstan that lowered the massive *ingegno*, manipulating it on the coral bank and raising it to retrieve the coral, a task that continued night and day. If there was no wind, the *corailleurs* also rowed the 16-ton boat. All of this was done under the searing North African sun (illus. 169).

The French could not compete with the lower production costs in Naples for boats and equipment. Lacaze-Duthiers therefore proposed 'Frenchifying the sailors and foreign boats' by a series of modern incentives (signing bonuses, military draft exemptions, family healthcare and so on), as well as suggesting measures to manage and protect the coral resource (such as fishing closure during the coral reproductive season, as was done in fin fish management). By 1870 almost all of the 160 boats fishing in Algerian waters were Italian, albeit under the French flag.[47] But by then Torre del Greco was solidly established in the 'complete coral cycle' of the wider Mediterranean.

Above: 169 Late 19th-century painting of coral fishers deploying a large ingegno or St Andrew's Cross. Opposite page: 170 Ceiling painting in the Nishidomari Tenmangu Shrine showing Daikoku, one of the seven Shinto deities (of Hindu origin) of fortune and a symbol of wealth, buying a magnificent colony of red coral.

Between 1875 and 1880, three deposits were discovered near Sciacca, Sicily. These consisted of mostly dead but still usable and readily gatherable coral of a colour previously unknown in the Mediterranean. A 'red gold rush' of two thousand boats produced a market glut and eventual decline in demand, when the price of even the best raw coral fell by 95 per cent, which in turn led to a precipitous drop in the number of boats fishing. The mostly family-owned 'coral firms from Torre del Greco collected as much

as eleven million kilo[grams] – a mountain of coral.'⁴⁸ 'Ironically, the Sciacca coral that drove prices down in 1880 is the most valuable *C. rubrum* variety today.'⁴⁹

Assocoral, the Italian trade association for coral and cameo manufacture and commerce based in Torre del Greco, says that there are about three hundred companies in the industry (including many subcontractors who finish the market product), employing nearly four thousand people. By now, most of the coral comes from company stockpiles or the Pacific fishery.

Italian companies were early traders in high-quality precious coral from Japan, which began to arrive in Italy in the first years of the twentieth century. Today most of the precious coral fished comes from Japan and Taiwan, and most of it is worked in Italy. The larger colony size of Pacific coral species allowed more elaborate sculptures to be created by Italian and East Asian artists.

In Meiji-era Japan, the coral business initially involved personal negotiations between coral fishermen and merchants, who might be boat owners, coral processors or shop owners. This interaction changed to a bidding system, eventually including Italian as well as Japanese merchants.⁵⁰ Nowadays a single specimen can fetch tens of millions of yen (U.S.$85,000) and with such sums involved, the wives and families of the sellers are present to 'make sure that . . . their husband or father doesn't spend it all partying before he gets home'.⁵¹ (illus. 170)

Thus, in Japan as around the Mediterranean, individual families ran the coral business, affording flexibility in company decision-making and negotiations. Eventually came the development of collaborative organizations of multiple merchants seeking to stabilize the industry and protect their profits.

Preserving Precious Corals

Its long-time use in the businesses of jewellery, luxe art objects and aquarium adornment, and questionable inclusion in some naturopathic nostrums, so depleted Mediterranean red coral that it would take centuries to recover even a fraction of its former abundance and the large colony size essential to maximum fecundity. The coral advocate Andrew Bruckner has likened the effects of the expanding fishery 'to that of strip mining and clear cutting of forests',[52] as Marsigli had recognized three centuries earlier. In North Africa (where legislative protection of the coral resource has a long history), the damaging European *salabre* was banned in 1823 in a treaty signed by France and the Bey of Tunis.

Sharp declines in the catch during the early 1980s and recognition of the dire condition of Mediterranean red coral populations prompted the General Fisheries Commission of the United Nations Food and Agriculture Organization (FAO) to sponsor international discussions that improved management and fishing practices for the Mediterranean. These measures included the forced abandonment by the EU in 1994 of the St Andrew's Cross and the *barra italiana*, already banned earlier in Spain because of its destructive effects in the Alborean Sea between Spain and Morocco.[53] But according to Bruckner, increasing catches beginning in the late 1990s reflected the growing take from previously unexploited deep populations, not a widespread recovery of the species. There remains rampant poaching by scuba divers at shallow depths.

Several unsuccessful attempts were made to list *Corallium rubrum* as well as Pacific species of *Corallium* and *Paracorallium* for protection under the Convention on International Trade in Endangered Species (CITES). There remains no consensus that even the decimated populations in the Mediterranean are sufficiently at risk to warrant a CITES listing. CITES regulations are not substitutes for fishery management nor do they ban exploitation, but only regulate trade. The jewellery industry worries that a stigmatizing effect of a CITES listing would imply a 'non-green' business and thus depress sales, harming coral artisans and small family businesses.[54] Because precious corals have both a high market value and the lowest growth and renewal rates of any commercially fished species, both trade control and fisheries management approaches are advisable.[55]

Like the centuries-old pattern in the Mediterranean, destructive dragging for precious corals in the Pacific led to a cyclic 'boom-and-bust' industry: discovery of new coral banks saw their excessive exploitation followed by their depletion, decline in profitability and desertion for the next virgin site. The inshore coral fishery in Japan has been managed locally since it began in Kōchi Prefecture in the 1870s. The boom-and-bust pattern typical of overfishing led to restricting the number of licensed fishing vessels, which since 2012 must be non-motorized. In the two legal fishing areas the catch is regulated, and fishing is closed during the annual season of coral reproduction and larval settlement. In Kagoshima and Okinawa Prefectures, the only fishing is selective, by submersibles, the high price of which deters newcomers from entering the industry.

Around Okinawa in the Ryukyu Islands, ecological studies pointed to selective collecting of large specimens of *Paracorallium japonicum* as one route towards a sustainable fishery, including a 'rotational harvest' with a 'biological rest period' of ten to twenty years that would allow coral regrowth.[56] The optimistic suggestion is admirable, but such patience historically has not been the watchword of the industry. Recall Count Marsigli at the turn of the eighteenth century advising that renewal of the 'forests' of precious coral required several decades. Fishers continued their efforts regardless until it was impossible to find corals of usable size: 'Assuredly that deficit comes not from nature but from not allowing time for its increase.' If the needed patience is not economically feasible, an alternative is to increase the size of marine protected areas and their proximity to fishing areas to promote larval recruitment into depleted populations as part of a management strategy that includes minimum size limits.[57]

The restraint of coral fishing seasons punctuated by no-fishing periods in coastal home waters of Japan and Taiwan did not extend to international waters, where coral was taken without pause until it became unprofitable. The two species of 'Midway corals' were fished relentlessly from the mid-1960s until the late 1980s. The devastating biological and subsequent economic effects led to the fishery being abandoned.

The late Richard W. Grigg, an expert on the biology and management of precious corals, related how colleagues at the University of Hawaii found plentiful precious pink coral off Oahu, but recognized the destructive nature of their dragnet sampling adopted by the fishers who followed them. Selective collecting by a manned submersible would be necessary for a sustainable fishery. This experience led to a forty-year research programme at the University of Hawaii on the management of precious corals.[58]

Black antipatharian corals also have been used in making contemporary jewellery, and Pacific black corals generally occur at shallower depths than red and pink octocorals (illus. 171). The discovery in 1958 of beds of black coral (*Antipathes*) at depths of 35–100 m (115–330 ft) off the island of Maui led to the development of a new, sustainable precious coral industry, in collaboration with the University of Hawaii's Sea Grant Program. The manned submersible

171 A diver about to take a small colony of Antipathes dichotoma *(now* A. griggi*) 50 m (164 ft) deep off Maui, Hawaiian Islands.*

Star II figured prominently in this selective taking of only the largest colonies, with some areas being set aside for scientific research and as possible reproductive refugia.[59] There has been some exploitation of the zoanthidean gold coral, *Kulamanamana haumeaae*, but discovery of its very low growth rate and great longevity prompted a moratorium (illus. 172).

The worry that the perceived rarity (evinced by a CITES listing) of precious corals might increase market demand and prices, inciting poaching and illegal trade,[60] and thereby cause management to fail, seems to have been justified. After designating precious corals as protected species in its waters, thereby restricting domestic fishing, in 2008 China used a CITES provision to seek the cooperation of other CITES parties in controlling trade in four *Corallium* species not deemed to be endangered globally. One effect of protecting precious corals locally was to raise their value dramatically in China, with its growing number of wealthy consumers desiring this ancient symbol of health, prosperity and success. In Taiwan, the major purveyor of precious coral souvenirs to Chinese tourists bent on conspicuous consumption, the price of the raw material jumped from U.S.$900/kg in 2009 to $7,500/kg in 2014, in step with an identical increase in Chinese tourism.

Unable to drag for coral legally at home, Chinese coral captains and crews began poaching in the territorial waters and exclusive economic zone (EEZ) of Japan. Poaching reached crisis proportions in 2014, when about two hundred large Chinese vessels, many flying their national flag, were fishing off the Ogasawara and Izu islands. Nozomu Iwasaki, an expert in the ecology and conservation of precious Pacific corals, argued that because of China's CITES stance in seeking international protection of precious corals in its waters, China would have a hard time rejecting Japan's request for protection from poaching in Japanese territorial waters and EEZ.[61] A summit meeting between the Japanese prime minister Shinzō Abe and Chinese president Xi Jinping defused the situation when Xi recalled the poachers. Subsequently, Chinese captains in violation were arrested and their vessels seized, apparently without official repercussion from China. Incidences of poaching had declined by early 2015, after more arrests and increases in the fines for poaching.

The perception of precious corals as luxury goods favoured by a wealthy elite (modern-day incarnations of Rossetti's *Monna Vanna*) seems to conservationists a threat to *Corallium* and *Paracorallium* species. But human exploitation in a new staging of the tragedy of the commons pales in the light of the potential effects of global climate change, which at the least may compromise any chance of recovery by the decimated coral populations. Increasing awareness of unregulated coral fishing and trade, and the threat posed by climate change, inspired the non-profit marine conservation organization SeaWeb's Too Precious to Wear campaign, launched in 2008.[62] Some jewellery buyers, designers and manufacturers (especially independent artisans and small businesses, but also Tiffany & Co.) have pledged not to use or buy precious corals lacking documentation as having been obtained sustainably, and some eschew coral entirely.

Opposite page: 172 The Star II *submersible returns from a dive off Oahu, Hawaiian Islands, with a haul of coral, including the gold coral* Kulamanamana haumeaae.

6 Coral Construction

> With Pantheist energy of will
> The little craftsman of the Coral Sea
> Strenuous in the blue abyss,
> Up-builds his marvelous gallery
>   And long arcade,
> Erections freaked with many a fringe
>   Of marble garlandry,
> Evincing what a worm can do.
>
> Laborious in a shallower wave,
>   Advanced in kindred art,
> A prouder agent proved Pan's might
> When Venice rose in reefs of palaces.
>   Herman Melville, 'Venice' (1891) (illus. 173 and 174)

Among the early evocations of 'coral architecture' was Carsten Niebuhr's account of the 'astonishing mass of the works built by marine insects' that he observed along the Red Sea coast while on a Danish expedition in 1762.[1] The accretive and modular growth of coral reefs, and the discovery of the animality of coral polyps, inspired the Victorian vision of these 'little craftsmen' as an industrious society of architects and builders. Their massive constructions were seen as undersea castles and palaces, even cities and continents, recalling the lost Atlantis. More prosaically, proximity and practical necessity saw the use of fossil and living coral rock as a building material by indigenous and colonial cultures from the Red Sea to Polynesia, the Caribbean and the Gulf of Mexico. During the Second World War, in the vastness of the Pacific, runways built of crushed corals were crucial in the Allied strategy of island hopping towards the Japanese homeland. Today, at a time of new tension in the South China Sea more than 1,400 km (850 mi.) from its mainland's shore, China is dumping coral sand and rock atop reefs that were once underwater to create islands and airfields where there were none. The new islands have killed the coral reefs that are their foundations, and their potential militarization has profound geopolitical implications in a sea crossed by global maritime trade routes.

Opposite page (top): 173 Claude Monet, Palazzo da Mula, Venice, *1908, oil on canvas; (below) 174 Zarh Pritchard,* Coral Arches, *1930, oil on leather.*

Coral Architects

The origin of the conceit of coral animalcules as architects may rest with descriptions of reef growth such as that by Matthew Flinders, who did not use the word 'architect' in *A Voyage to Terra Australis* (1814). But his telling, like that of Johann Reinhold Forster (a naturalist on Cook's second Pacific voyage), turned teleological and credited the coral polyps with foresight. Flinders's description carried the picturesque notion of coral architecture, with the polyps building a reef that became a 'monument to their wonderful labours'.

Charles Darwin himself, on 3 April 1836, diarized while on his *Beagle* voyage:

> We feel surprise when travellers tell us of the vast dimensions of the Pyramids and other great ruins, but how utterly insignificant are the greatest of these, when compared to these mountains of stone accumulated by the agency of various minute and tender animals!

Darwin had read Flinders, and in a subsequent entry on 12 April he borrowed the metaphor and referred explicitly to the 'myriad of tiny architects' responsible for building the monument that is a lagoon island. More mystically, Samuel Taylor Coleridge in his contemplation *Hints Towards the Formation of a More Comprehensive Theory of Life* (published posthumously in 1848) had seen the growing islands of Polynesia as a 'gigantic monument, not so much of their [the polyps'] life, as of the life of Nature in them'.

In 1824 John MacCulloch, in the *Boston Journal of Science*, attributed communal reef building by the scattered, seemingly unconnected polyps to a 'mysterious principle' akin to what today would be called a 'hive mind', in discerning the colonial, even cultural, nature of corals.

Later, the anthropologist Alfred L. Kroeber would focus on the social element:

> Such a reef may be miles long and inhabited by billions of tiny polyp animals. The firm, solid part of the reef consists of calcium carbonate produced by the secretions of these animals over thousands of years – a product at once cumulative and communal and therefore social . . . Each of us undoubtedly contributes something to the slowly but ever changing culture in which we live, as each coral contributes his gram or two of lime to the Great Barrier Reef.[2]

R. M. Ballantyne, too, cast the coral 'insects' as architects in a social, moral and religious framework in *The Coral Island* (1858). Ballantyne saw himself as an educator and imparted information about natural history and theology, as well as ethnography of the South Pacific. He put polyps into a divine plan, in which 'we consider . . . the smallness of the architects used by our heavenly Father in order to form those lovely and innumerable islands' as part of His manifold works.

Ballantyne's book was but one of many (for example, those of Philip Henry Gosse) at that time mingling science with evangelical Christianity. Moreover,

> the coral insect captured the Victorian imagination as the epitome of industriousness, reflecting contemporary belief in the value of labor and production . . . Coral figures the sublimation of the individual to the social, rendering what is ordinary, small, and unprepossessing, into a . . . structure as magnificent as a coral island, and thus for the early Victorians the synergetic superiority of coral signified the divine order in an age that was still able to see science as compatible with, indeed a witness to, the power of God.[3]

Another publication at the time seemingly promulgated the Darwinian subsidence theory of atoll formation over the 'elevated volcano' hypothesis, but cast the polyps as God's building contractors:

> When the earth's crust is collapsing, and it becomes necessary to fill up the vacancy, the commission is not given to any gigantic workmen, but a number of mere polyps are bid to labour upon the subsiding soil, as if to show that the Creator could employ the humblest of His creatures in executing the largest of physical undertakings.[4]

6 CORAL CONSTRUCTION

In *The Sea* (1864), Jules Michelet rhapsodized about the accretive capacity of corals as 'The World Makers'. Seeing Lamarck's madrepores in the Muséum National d'Histoire Naturelle in Paris, he gave them a voice:

> Time – give us only time, and these rocks will become hospitable, tenanted, fruitful . . . [and] will no longer have these terrible threatenings for the seaman. We are preparing a spare world to replace your old one should it perish . . .

Dr Jules Rengade (under the nom de plume Aristide Roger), in *Voyage sous les flots* (Voyage beneath the Waves) of 1868, would reiterate Michelet:

> This forest of calcareous concretions, the pillars of marble, the columns of coral, the pedestals, the madrepores and polyparies were the base of a future land, of a world to come. The diaphanous and gelatinous polyp, at once the architect and builder of these immense undersea works, was preparing under the veil of the waters a new land, for beings more perfect than itself.

In 1848, Charles Dickens had written in a book review that 'shining cities glittering at the bottom of quiet seas . . . exist no longer; but in their place, Science, their destroyer, shows us whole coasts of coral reef constructed by the labours of minute creatures . . . that have passed away.'[5] Passages such as this, as well as Michelet's book and those of his British contemporary Gosse and others such as Matthias Jakob Schleiden's *Das Meer* (The Sea, 1867), reflected the profound change in how corals had come to be regarded by the mid-nineteenth century – as wonders of creation and creators themselves.

Coral colonies and reefs became castles and fortifications. Early on, in his narrative of the voyage of the *Fly*, Joseph Beete Jukes remarked on the structural geology of the Great Barrier Reef and likened it 'to a gigantic and irregular fortification, a steep glacis crowned with a broken parapet wall . . . The tower-like bastions, of projecting and detached reefs, would increase this resemblance.' C. M. Yonge, in his memoir *A Year on the Great Barrier Reef* (1930), described 'little patches of *Galaxea*, fields of fairy castles each crowned with a battlement of spines' (illus. 175 and 176). Melville, in *Omoo*, wrote of 'the coral rampart', and Umberto Eco in *The Island of the Day Before* also related reef to castle when he wrote of the rich fauna inhabiting its 'barbican'.

Jules Rengade provided some of the most explicit fictional prose. Although his reefs beneath the Coral Sea contained the usual array of zoophytes, lithophytes, alcyonaceans and gorgonians found in scientific books of the time, and blossomed with standard botanical imagery, the polyps' prowess as both architects and builders dominated. Like Darwin, Rengade used the monumental pyramids for scale:

> The dimensions of these colossal pillars were incalculable. Their base, as wide as that of a pyramid, rested on the entire surface of an underwater plateau; their cavernous flanks, sculpted like a medieval church's steeple, comprised crowds of curious animals . . .
>
> The pillars and columns took on more and more colossal dimensions; the arches, porches, balustrades and flying buttresses multiplied themselves infinitely and intertwined in a thousand picturesque fashions; the gigantic polyps' boughs fused to form vast porticoes, and gradually this immense forest of madrepores was turning into a magic palace.[6]

Although describing craggy Edinburgh, Robert Louis Stevenson (who knew of coral reefs from Ballantyne's book and would later visit them)

might have been envisioning such a scene when he wrote of a 'dream in masonry and living rock'.[7]

Writing two years after Rengade, Edgar Quinet (in *La Création*) reversed the simile, when, on a post-human earth, our triumphant successors would regard our geometry as we admire the bee's instinctive hexagons, and might view the Parthenon as a lovely coral reef ('*banc des polypiers*'). H. G. Wells likewise took his readers in *The War of the Worlds* (1898) to an almost post-human earth and compared the ruins of a London destroyed by invading Martians to the 'human reef' it once was. (The Martians had fled their own dying planet to colonize another that was in the early throes of an industrial revolution and setting out on its own path to catastrophe.)

Rengade's conjuring of coral concretions as pillared and porticoed palaces on an oceanic plateau coincided with the re-emergence of Atlantis in the nineteenth-century Romantic imagination. The illustration at the start of his chapter 'The Coral Sea' ambiguously showed a coral reef as classical ruins being reclaimed by an undersea jungle of zoophytes, or as an ongoing construction project clad in biomorphic scaffolding (illus. 177). The latter image recurred verbally in Peter F. Hamilton's science fiction novel *Pandora's Star* (2004): on a distant planet

Previous page: 175 'Little patches of Galaxea, *fields of fairy castles each crowned with a battlement of spines.' Above left: 176* Galaxea fascicularis *(fig. 1, detail of plate 55 in John Ellis,* The Natural History of Many Curious and Uncommon Zoophytes *(1786)); skeleton of G.* fascicularis. *Above right: 177 The engraving of a submarine scene at the start of the chapter 'La mer de corail' (The Coral Sea) in Jules Rengade [Aristide Roger],* Voyage sous les flots *(*Voyage beneath the Waves; *1869), shows coral constructions as classical architecture.*

colonists use a genetically engineered extra-terrestrial organism, 'drycoral', to build houses by allowing it to envelop a framework with a pumice-like organic shell.

In the middle portion of her book *The Artificial Kingdom* (1998), Celeste Olalquiaga expounded at length on the melding of Atlantean ruins and corals, via another nineteenth-century fascination, the marine aquarium. Aquaria incorporated Romantic elements of fantasy gardens and artificial ruins as well as geological grottoes. Among the foremost inhabitants of Victorian aquaria were corals, which while appearing inert actually were accreting and creating 'fantastic underwater architectures'.

And for Michelet (who had collaborated with Quinet),

The stone was not simply the base and shelter of this people; it was itself a previous people, an anterior generation, which, gradually overtopped by the younger, assumed its present consistence. And all the movements of that first community are still strikingly visible, as details of another Herculaneum, or Pompeii.[8]

The image of living reefs building on an anterior generation is apt. Reef corals grow by accretion, with new skeleton being deposited on the surface of the colony, atop older layers that they preserve unchanged. The simile was also used by Steve Jones in *Coral: A Pessimist in Paradise* (2007): 'Like Rome, most of today's reefs rest upon the ruins of their past and, as in that city, their remnants reflect the ebb and flow of history,' with an added palaeontological perspective, for 'viewed across the abyss of time, reefs come and go.'

Olalquiaga continued: 'The constant interchangeability between coral, rocks, and ruins paved the way for their final allegorical mutation, making the ocean floor an artificial kingdom that surpassed in the nineteenth century all legendary descriptions of Atlantis.' The view of coral reefs as classical architecture persisted in poetry. John Wheeler, inspired by the Great Barrier Reef, in his sonnet 'The Coral Reef' (which Horace Keats set for piano and voice; see Chapter Four), saw coral architecture as shadowy temples, towers and colonnades in 'Em'rald glades'.

James Hamilton-Paterson, in *Playing with Water* (1987), looked down on the corals' underwater city and saw 'squares and minarets, arcades and loggias'.[9] Later, in *The Great Deep*, while diving at night on a coral reef, he mused, 'The huge unseen city itself seems always on the cusp of vanishing, it is so delicate and its true nature so elusive.'[10]

But if fabulous cities could disappear into the sea, so too might they emerge from it. Mindful of his earlier years among the coral reefs of the South Pacific, later in life Melville published a collection of poems, 'Fruit of Travel Long Ago', in the volume *Timoleon, Etc.* (1891). The collection included 'Venice' (quoted in full as the epigraph for this chapter), which 'rose in reefs of palaces'. Ironically, in a warming globe with rising sea levels, Venice, like Atlantis before it, is now drowning.

Cinematic vignettes also conflated corals with Atlantis. Georges Méliès propelled Yves, the dreaming fisherman and submariner, past coral formations backed by Atlantean ruins (and a sunken sailing ship) in his surreal film *Deux cent milles [lieues] sous les mers, ou le cauchemar du pêcheur* (1907).[11] The James Bond film *For Your Eyes Only* (1981) picturesquely submerged Aegean ruins amid mostly soft corals in Bahamian waters, but the setting was essentially an exotic backdrop of classical clichés.

If writers and film-makers looked backward and saw coral reefs as sunken cities and lost civilizations, the Paris-based conceptual architect Vincent Callebaut looked hopefully towards the human future. Mindful of reefs as cities that succeed through symbiotic exchanges, the theme of urban recycling informed his response to the earthquake that devastated Haiti in 2010. Inspired by the organic form, modular construction and sustainability of a coral reef, he envisioned a futuristic coastal complex built from modules to house one thousand Haitian families in a 'sustainable urbanity'.[12] (illus. 178) Roofs of the modules would have vegetable gardens fertilized with recycled wastes, and there would be ponds for aquaculture of fishes, with their waste also recycled. Like a coral reef, the village would be solar-powered (by photovoltaic panels), with additional electricity from wind and underwater turbines to harness air and sea currents, as well as the capture of the potential energy in the thermal difference between shallow and deep seawater. In his 'Neo-nature' project, the interaction designer Michail Vanis envisions using reefs of engineered

corals themselves to capture and direct currents towards electric generators.[13]

Coral in Construction

The durable structures built by corals, and practical necessity, no doubt both figured in the adoption of coral as a construction material by humans. Well aware of the nature of coral and the structure of coral reefs, William Saville-Kent punned in the preface to his monumental *The Great Barrier Reef of Australia* (1893), where he promoted his pioneering photographs 'for the delineation of coral-reef structures in the concrete'.

Cutting slabs from terrestrial deposits of fossilized coral is one thing, but chiselling and prying living 'coral rock' from a reef is quite something else. Both materials have been used in construction, with varying extents of environmental consequences. The Kuna people

Above: 178 As a response to the devastating 2010 earthquake in Haiti, the Coral Reef Project by Vincent Callebaut Architectures, mindful of the wondrous architecture of coral reefs that succeed through solar-powered symbiosis and recycling, envisions a three-dimensional and energy-self-sufficient village built from a standard, prefabricated module in order to rehouse the refugees from such humanitarian catastrophes. Opposite page: 179 Engraving based on a sketch by W. Wilson of the stepped ahu of Marae Mahaiatea on Tahiti, measured by Cook and Banks in 1769.

have inhabited some of the 350 small islands of the San Blas archipelago off Panama for about 150 years. Traditionally, the Kuna have mined living coral rock to reinforce with sea walls the eroding margins of their islands, and even to enlarge the islands and build new ones. Today, with the population growing and sea level rising, the reefs are diminished, overfished, and fouled by algae, but the electrolytic accretion technology of Biorock™ is being used to create new substrata for reef growth as breakwaters and as habitat for fishes and lobsters.[14]

On coral islands in the Pacific and elsewhere, it was natural to use coral in construction intended to be permanent. In the Society Islands (including Tahiti), coral rock from the reef was incorporated into religious and ceremonial structures called *marae*. These had a large, rectangular, paved open area or court, with a raised altar platform (*ahu*) at one end. James Cook and Joseph Banks measured one such *ahu* on the Marae Mahaiatea that had been built just before their visit in 1769, when Tahiti was warring with Moʻorea (illus. 179). This *ahu*, 81.4 by 21.6 m (267 × 71 ft) at its base, had eleven stepped terraces, each 1.2 m (4 ft) high, with solid blocks of dressed coral up to 0.8 by 1.1 m (2.5 × 3.5 ft) forming the edges of the steps.[15]

Absolute dating of the *marae* has been enabled by the precise thorium–uranium (^{230}Th/^{234}U) technique, which in conjunction with radiocarbon dating allowed a chronology of *marae* on Moʻorea. Both branching *Acropora* and massive *Porites* species were incorporated into *marae*, and the excellent condition of the delicate surface features of *Acropora* skeletons indicates that they were cut from the reef while alive. The progression from simple to complex ritual architectures between the fifteenth and

GREAT MORAI of TEMARRE at PAPPARA in OTAHEITE.

seventeenth centuries was concomitant with the ascension of the Oro war cult and its emphasis on sacrifice.[16] Perhaps tellingly in this sacrifice of living beings, fresh corals were brought from the coast to inland *marae* as ritual offerings and not just as construction material.

In his book on the civilization of ancient Hawaii, *A Shark Going Inland Is my Chief* (2012), Patrick Kirch tells us that far to the north of French Polynesia, when temples proliferated during the sixteenth and seventeenth centuries, living coral was also offered to the gods of sailing canoes and of fishes at small coastal shrines maintained by fishermen. Further inland on Maui, coral branches were presented on the altars of larger temples (*heiau*) associated with Kū (the war god), Kāne (associated with procreation and irrigated agriculture) and Lono (god of rainfall, agriculture and sweet potatoes). Participants in a rite at a king's war temple purified themselves in a sea bath and then carried branches of coral to pile up at the *heiau*.[17] Such offerings suggest an early awareness by Oceanic peoples of the living or even animal nature of coral: it seems more than coincidental that offerings of corals – living beings having the form of plants but emitting the smell of decaying animal flesh when dead (recall the olfactory accounts by Rumphius, Peyssonnel and Jukes) – were made to those gods that would assure a successful agricultural cycle, *and* to a deity linked with death.

Coral also appeared in ritual caches dating to 300–600 CE (early Classic period) at Tikal, the quintessential Maya city and ceremonial centre in Guatemala. Razor-sharp branching corals might have been used for the bloodletting well known in Maya ritual, or as a source of toxins, or simply for their marine symbolism.[18] Some caches held mostly brain corals, which are neither particularly incisive nor envenoming to humans, so symbolism was their more likely purpose – perhaps a cerebral memento mori in a culture steeped in human sacrifice and the display of skulls.

Buildings at inland Maya sites such as Tikal were made of locally abundant limestone. The same was true in Belize, where limestone and sandstone were quarried for buildings on the mainland (some of which notoriously have recently been destroyed and used to build roads). Coastal Maya communities on Belizean cays in the post-Classic period, 900–1500 CE, laid foundations from massive corals laboriously quarried from the reef, rather than building directly on the ground.[19] The coral foundations (many today are partially submerged in the 'drowned landscapes' of rising sea level) were faced with quarried limestone brought from the mainland. Fragments of 'finger corals' were poured over the boulder foundation to produce a level sub-floor. Traditionally, the Maya interred deceased family members in these foundations, beneath the floors of dwellings, mingling the bones of humans and corals.

Pacific peoples quarried coral megaliths for monumental and ceremonial constructions such as the Taputapuatea complex on Ra'iatea in the Society Islands, the cradle of Polynesian culture dating perhaps to the tenth century. There, the *marae* Hauvivi 'was enclosed to the height of six feet seven inches by a wall of gigantic coral blocks . . . hewn from the inner reef' (illus. 180).[20] Since the sixteenth century the site has been consecrated to Oro, the god of war and fertility. Elsewhere, on Tongatapu in Tonga, the Ha'amonga'a Maui (Maui's Burden, named after that hero) is a monumental trilithon that likely was built as the entrance to a royal compound, or, more controversially, as an astronomical observatory (illus. 182).[21] Less questionable in their interpretation are the royal tombs, *langi* (also dating to the thirteenth century), built from huge slabs of coral brought by canoe from nearby islands.

Not all Oceanic peoples interred their dead in monumental tombs such as those on Tongatapu. In some cultures, the bodies, skulls

180 Coral blocks nearly 2 m (7 ft) tall surround Marae Hauvivi in the Taputapuatea marae complex on Raʻiatea, seat of the Oro cult.

in particular, were preserved in shrines, often in family homes. On Nusa Kunda, or Skull Island, in the Solomon Islands, a communal shrine built of coral slabs housed skulls moved there to protect them from destruction by Christian missionaries (illus. 181).[22] Although to the susceptible outsider this construction evokes the link between coral and death, such was likely not the intent of the builders, who used the most durable building material available to house reverently their forebears' remains.

Not only the original Polynesian colonists but ensuing ecumenical immigrants, too, used coral rock in their religious edifices. After

Kamehameha I opened Hawaii to *haoles*, Congregationalist missionaries in what is now Honolulu began building the so-called Stone Church of 450-kg (1,000-lb) coral slabs hand-chiselled by native Hawaiians from several metres' depth on the living reef. According to a local historian, some 14,000 such slabs were taken to shore on canoes and used to construct Kawaiaha'o Church, which was dedicated in 1842.[23] A year later, the Catholic Cathedral of Our Lady of Peace, also built of coral blocks quarried nearby by the islanders, was consecrated.

Three centuries earlier, Caribbean reefs had provided the blocks for Christopher Columbus's successors to build the Catedral de Santa María la Menor in Santo Domingo (completed in 1540), in today's Dominican Republic (illus. 183). Francis Drake used the cathedral as his headquarters when he captured the city in 1586. The same sort of material was used in Cuba for Havana's baroque Catedral de la Virgen María, completed in 1777.

The littoral of the Indian Ocean is littered with the remains of buildings constructed from coral. Some of the best-preserved are on the United Nations World Heritage list of culturally important sites. Stone Town in Zanzibar, Tanzania, with its 1,700 buildings of coral (some dating perhaps to the twelfth century), stands at the centre of an archipelago of such trading towns on the Swahili coast of East Africa. The Swahili (Arabic for coastal dweller) trading culture dates

Below: 181 Shrine on Nusa Kunda, or Skull Island, in the Solomon Islands, with skulls in niches on the platform of coral rock. Opposite page: 182 Ha'amonga'a Maui, 13th century. Front view of Remarkable Stones; side view of Remarkable Stone. Photographs c. 1880–89.

Front View of Remarkable Stones   191.

Side View of Remarkable Stone   178.

to about 800 CE (a relative latecomer to the newly appreciated maritime Silk Road of East–West trade) and looked outward, towards the Indian Ocean.[24] Mosques are prominent in these towns, particularly ancient Lamu in Tanzania, where their construction (faced with white plaster) recalls that in Red Sea towns. Further east, in the coral archipelago nation of the Maldives, the original Buddhist culture used coral in architecture and sculpture, which subsequently became more refined, particularly the carvings in the Islamic constructions of tombs and mosques.

The eighteenth-century German explorer Carsten Niebuhr noted that coral rock living underwater was soft enough to be sawed easily and was a favourite material for building houses all along the shore of the Red Sea. He related that for his late companion Peter Forsskål (who studied with Linnaeus), each of these Arab houses constituted a 'natural history cabinet' as rich in corals as any collection in Europe. Already blocks of hard coral had been used in the Roman era at Berenike on the Red Sea to build quays and jetties, private houses and even a temple to Serapis.[25]

Dwellings built of coral rock had utilitarian, humble beginnings, albeit not without their aesthetic touches. 'Like the dhows which anchor in the shallows over the coral reefs . . . the buildings seen across the water seem to hang suspended (illus. 184).' Thus did the architect Derek Matthews begin his 1953 article 'The Red Sea Style' about the coral buildings of Suakin, a once-thriving port on the Sudanese coast of the Red Sea dating to Egypt's New Kingdom, in the Land of Punt. Colonel Kitchener had his headquarters in what since the sixteenth century had been Ottoman Suakin, and General Gordon stopped there en route to his doom in Khartoum during the 1885 insurrection of the Sudanese Mahdists.

183 *Cathedral of Santa María la Menor, Santo Domingo, the oldest cathedral in the Americas, built 1514–35.*

But the very security of the port of Suakin from attack by the sea – its narrow approaches are defended by coral reefs – made it dangerous for large, modern commercial vessels, leading in 1909 to the opening of new facilities further up the coast that became Port Sudan. The ancient city of Suakin became a tumbledown ghost town (illus. 185). Its twin town on the opposite shore, Jeddah, thrived by building seaward, atop its coral reefs. A century later, after the lifting of U.S. sanctions against Sudan, Turkey's President Erdoğan announced in December 2017 that Sudan had given permission to reconstruct Suakin Island as a cultural waypoint for Turkish pilgrims en route to Mecca via Jeddah.[26] This rebuilding, with the addition of a civilian and naval docking facility, may resituate Suakin in a new world order, with unknown consequences for its coral reefs and lagoon.

The porous, friable coral rock cut from the reef and used in the construction of houses and mosques in Suakin and other Red Sea towns was a risky building material, especially using

Above: 184 Ernst Haeckel's watercolour of himself and the crew in a dhow above Red Sea coral reefs off Al-Tur (Tor), Sinai. From Ernst Haeckel, Arabische Korallen *(1876), plate v. Opposite page: 185 The ruins of the old port of Suakin rest atop a circular island of coral bedrock in the lagoon that provided a rare natural harbor on the Red Sea coast. Khorshid's house is the large single-storey structure set back from the shore at the bottom left of the aerial photo.*

impermanent mortar mixed using seawater, and walls had to be reinforced with scarce timber (imported to the Arabian Peninsula from Southeast Asia since the days of the *Periplus of the Erythrean Sea* in the first century CE) and covered with waterproofing plaster made from crushed coral.[27] The plaster was decorated with traditional Islamic arabesques and floral designs, predominantly in the houses' interiors. Most of Suakin's architectural ruins are beyond restoration or preservation. Instead there are plans to reconstruct certain of them from conserved original coral blocks, wood and casts of decorative details, using early drawings and photographs of them before they collapsed, as a living museum and cultural site.[28]

One of Suakin's oldest and finest permanent houses was that of Khorshid Effendi, built perhaps in the early sixteenth century using techniques brought from the Arabian peninsula, but altered often since then.[29] Surrounding the traditional solitary rosettes in the centre of the plastered spandrels of a now collapsed arch in a *diwan* (greeting room) fronting the channel to the sea was an unusual array of unconnected stars with six-part symmetry (illus. 188). This is not a familiar or traditional Islamic design, which most commonly are arabesques of interconnected eight- or six-pointed stars, or starry grids (as occurred elsewhere in Khorshid's house). The idiosyncratic motif resembles the separate corallites of a massive scleractinian reef coral, each having four cycles of hexamerous septa, but where a six-pointed star replaces the coral's central columella in the motif. Might a prosperous trader have decorated his home with a unique pattern of regularly spaced but unconnected starbursts inspired by the distinctive skeletal morphology of local coral, the most important building material available? As elsewhere, the motif recalls the 'starrystones' considered lucky for millennia, as discussed earlier. Its overarching position would have evoked the vault of the heavens themselves, familiar to early Islamic astronomers and later represented as the starry firmament in the famous wood engraving in

186 and 187 Walls constructed of coral in the old town of Jeddah, Saudi Arabia, were reinforced with timber such as this palm trunk (bottom), and the coral blocks were covered with plaster (top).

Above: 188 The plaster face of the spandrels of the arch in Khorshid's house is decorated with a stellate pattern that may represent corallites in the stone used to build the house. The pattern recalls the heavens and lucky starrystones.

Camille Flammarion's *L'Atmosphère: météorologie populaire* (The Atmosphere: Popular Meteorology) of 1888. The Latin words for 'star' (*astrum*, *sidus* and *stella*) have come to populate the galaxy of coral nomenclature.

Despite the importance of coral reefs as protective coastal barriers, especially under the present threat of rising sea level and more intense storms that may wash over them, the quarrying of coral for construction in Indian Ocean nations continues, albeit less than previously because of legislation and sustainable alternatives such as concrete blocks (which nevertheless require sand – often coral sand – in their manufacture), and the use of some of these alternatives as artificial substrata to encourage coral growth.[30] Steve

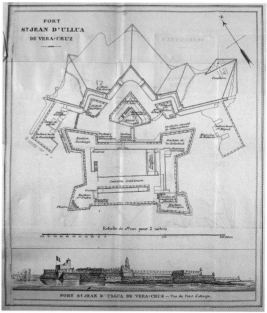

Opposite page: 189 Seaport of Jeddah, Saudi Arabia, and its coral reefs. Above: 190 San Juan de Ulúa, Veracruz, Mexico; right: 191 French plan and panorama of the fort from 1838, when Admiral Charles Baudin commanded the French forces at the Battle of Veracruz.

Jones had it right when he likened coral reefs to ancient cities built on 'the ruins of their past'. Henry David Thoreau, writing in *Walden* (1854), abhorred such places: 'Deliver me from a city built on the site of a more ancient city, whose materials are ruins, whose gardens cemeteries. The soil is blanched and accursed there.' Less metaphorically than in Jones, some human cities are built on or of the ruins of coral reefs (illus. 189).

In the 1950s the South Pacific Commission (now called the Pacific Community) used the historical precedent of Bermuda and Barbados to promote not only fossilized coral but coral sand and rubble (and burning coral reef rock to make slaked lime for mortar and cement blocks) as building materials. In Australia, the Queensland Cement and Lime Company burned coral that it dredged from Moreton Bay, off Brisbane, until 1995.

The solidity and proximity of coral of course occasioned its use in coastal fortifications. The wreck of his flagship *Santa María* on a coral reef off Hispaniola ended Christopher Columbus's first expedition to the New World in 1492. He returned to Hispaniola the next year and put his son Bartholomew in charge of a new colony, which the son eventually moved to what became the site of Santo Domingo. Indigenous insurrections and tropical hurricanes necessitated fortifying and strengthening the colony, and in 1503 a new governor began constructing the massive Ozama Fortress largely of coral blocks.

Hernán Cortés founded Veracruz, now Mexico's largest port, in the southwestern-most corner of the Gulf of Mexico, in 1519. Just offshore, many ships were wrecked or grounded in the Veracruz reef system, continuing into modern times. As the principal site for exporting gold, silver and spices, the wealthy port became a likely target for attack by pirates, including Elizabeth I's favourite, a youthful Francis Drake, defeated in a sea battle there in 1568 following a year of privateering in the Caribbean. Construction of the fort San Juan de Ulúa, begun in 1565 and expanded after this battle, continued through the eighteenth century. The immense structure was built entirely of *piedra de múcar* (coral rock) (illus. 190 and 191).

When Cortés arrived, there was nothing but sand dunes in the area, so coral was the only building material. Erecting this island fortification was not the only threat to nearby reefs:

> All public buildings of the 17th and 18th centuries, such as fortresses, hospitals, churches, government offices, the customhouse, and hotels, as well as small houses and the rampart that surrounded the old city of Veracruz, were built with piedra de múcar. There are no records of the quantity of coral heads used, but it surely included thousands of tons, as more than 1,000 structures were built with coralline rock . . . Ancient maps of Veracruz from the 17th and 18th centuries show three apparently well-developed reefs . . . all of which no longer exist.[31]

In *The Fatal Shore* (1987), Robert Hughes observed that 'Australia has many parking lots but few ruins.' Of the latter, the remains of a small 'fort' – the oldest structure built on that continent by Europeans – consist of coral slabs piled up by survivors of the 1629 wreck of the Dutch East Indiaman *Batavia* (illus. 192). The ruins rest on West Wallabi Island in the Houtman Abrolhos.

After the shipwreck, the survivors divided into two factions. The defenders' fort was built 'with a soldier's eye' on (relatively) high ground as protection from attack by the musket-bearing mutineers,[32] whom the defenders repelled by hurling a barrage of coral rocks. Two months later, a rescue mission captured the murderous mutineers and imprisoned them in jails built of coral slabs,[33] the next-oldest European buildings on the continent that would become Britain's penal repository.

As Kamehameha I consolidated his control of the Hawaiian archipelago at the turn of the nineteenth century, he welcomed foreign trading and whaling ships, accommodating their need for a mid-Pacific provisioning station and opening to them the favourable anchorages in the harbour at what would become Honolulu. In exchange for local products and a strategic position he received a now-familiar currency – armaments and military advisers – and proceeded to fortify the harbour. The fort's 340 by 300 ft (103.6 m × 91.4 m) long walls were built 12 ft (3.5 m) high and 20 ft (6 m) thick – of coral cut from the reef.[34] This construction would have required 11,378 cu. yd (8,699 cu. m) of coral rock. Barely four decades later, the walls were demolished and used to refill some of the harbour as downtown Honolulu expanded (illus. 193).

As warfare became modernized, the Second World War saw new uses for coral islands and atolls as stepping stones for the Americans and their allies island-hopping across vast Pacific distances towards the Japanese homeland. Coral rubble on the fringes of such islands, former reef flats buried under topsoil, or mining and dredging of the reef itself, provided a hard material for building and surfacing runways. A U.S. Navy civil engineer extolled the use of coral for such purposes:

> All-weather runways, parking areas, and roads . . . built on island after island by the

Opposite page: 192 The fort built of coral rock by survivors of the Batavia *shipwreck in 1629 on West Wallabi Island, Houtman's Abrolhos, Western Australia, is the oldest European structure on that continent. Above: 193 Paul Emmert,* View of the Honolulu Fort – Interior, *c. 1853, oil on canvas. Begun by Russians in 1815, the fort was finished by Kamehameha I. The height of this harbourside fortification can be judged from the ships' masts in the background.*

> Navy's Seabees . . . were possible only with coral. In fact, coral might well be called the world's best natural material for runway construction.[35]

The author did offer a caveat, however, regarding the use of 'hard coral underwater reefs that require heavy charges of dynamite before they can be handled with a backhoe or dragline'.

Another Navy officer noted the heavy wear and tear on construction equipment posed by the hardness of the coral, and that 'very hard coral "heads," which are classified as crystalline limestone', slowed operations.[36] Living coral often had insufficient cementing properties because of incidental washing during its blasting and dragging, but this shortcoming could be rectified by rolling and crushing the coral (illus. 194).[37]

Palmyra Atoll (now part of the Pacific Remote Islands Marine National Monument established by the U.S. in 2009 and considered one of the least-disturbed remaining reef systems) was a staging point for Second World War aircraft flying between there and Australia. Palmyra had two large runways (1,700 × 91 m/5,630 × 300 ft; and 1,100 × 61 m/3,600 × 200 ft) of crushed and rolled coral, as well as a landing site for seaplanes dredged in its lagoon. Tern Island (now the French

Frigate Shoals Airport) in Hawaii, originally only a few hundred feet long, was enlarged using dredged coral to accommodate the 945 by 85-m (3,100 × 275-ft) runway for fighter planes being shuttled from there to Midway Island. Appropriately, from above, the island-airport resembles nothing so much as the flight deck of an aircraft carrier (illus. 195).

Late in the war, Tinian (just north of Guam) in the Mariana Islands became the forward base for long-range aerial operations against the Japanese home islands. Building first on two existing Japanese structures, U.S. Navy Seabees constructed four parallel 2,590-m (8,500-ft) runways (with associated taxiways and hardstands) of rolled coral, which required moving 7.65 million cu. m (10 m cu. yd) of earth and coral; the compacted coral surfacing alone totalled 2.3 million cu. m (3 m cu. yd). This amount of coral dwarfs that consumed in the construction of the older coastal fortifications in Santo Domingo, Veracruz and Honolulu. When completed, the airfield on Tinian was the largest in the world and served more than 250 B-29 Superfortress bombers, the heaviest aeroplanes then used by the U.S. Two of these,

Opposite page (top): 194 Bulldozer crushing inconveniently large blocks of coral to be used in runway construction during World War II; (bottom) 195 French Frigate Shoals Airfield, 1961. Below: 196 Enola Gay on a hardstand of crushed coral at North Field, Tinian, in 1945.

Enola Gay and *Bockscar*, dropped the atomic bombs on Hiroshima and Nagasaki, respectively, in August 1945, to end the war (illus. 196).

More recently, during construction of the strategically important and politically controversial U.S. naval base on Diego Garcia in the Chagos Archipelago, 'equidistant from everything in the Indian Ocean', a private contractor dredged 1.9 million cu. m (2.5 m cu. yd) of coral from the reef adjacent to the island to accommodate submarines and large ships.[38] The original immensely sturdy 2,320-m (8,000-ft) runway (since enlarged and duplicated) was built of coral aggregate concrete designed to accommodate C-141 transport aircraft, and B-52 bombers have been based there. B-1B strategic bombers used the runway during the 2001 Afghanistan bombing campaign in the aftermath of the 9/11 terrorist attacks in the United States.

The Spratly Islands, an archipelago of more than six hundred islets, atolls and coral banks in the oil- and gas-rich South China Sea, sit astride shipping routes that carry trillions of dollars of trade to northeast Asia. Among six national claimants, the most determined is China, whose island-building atop coral reefs in the Spratlys beginning in 2013 raises political, diplomatic, economic, legal and military uncertainties and conjectures regarding China's policy of turning formerly international waters and airspace into its 'sovereign territory'.[39] The Chinese foreign ministry predictably discounts international fears, yet warns others away from its new sovereign territory where some islands bristle with radar and surface-to-air missile installations, and where a 3,125-m (10,250-ft) runway on Fiery Cross (Yongshu) Reef was inaugurated in January 2016. Six months later, in a claim brought by the Philippines, the Permanent Court of Arbitration at the International Court of Justice at The Hague rejected China's 'historical claims' in the area, a decision that China refused to accept. The U.S. military has tested the waters by warship cruises and aircraft overflights.

Opposite page: 197, 198 and 199 Fiery Cross (Yongshu) Reef in the Spratly Islands was largely under water in May 2014 (top) before the start of artificial island-building. Six months later (centre), suction dredgers had deposited coral sand to cover the northern side of the reef and extend it above high tide, also dredging a harbour and cutting a channel to it from the sea; dredging ships are visible in the harbour, as are the white plumes of sand in the water and long conduits through which sand and pulverized coral was pumped and deposited atop the reef. By April 2016 (bottom) Yongshu Island was capped by dredged coral debris and concrete, the foundation of a new Chinese base, including the 3-km (nearly 2-mi.) runway along the length of its northern edge that opened in January 2016.

But beyond the geopolitics lie the perturbing effects on one of the world's richest coral reef ecosystems, on one edge of the Coral Triangle. Between September 2013 and June 2014 alone, Chinese cutter suction dredgers dumped more than 10 million cu. m (13 m cu. yd) of sand onto reefs that were formerly underwater at high tide, to create artificial islands. The commander of the U.S. Pacific Fleet called this man-made landform China's 'great wall of sand'.

According to one assessment based on satellite photos, 'Coral reefs that have been left untouched for centuries by virtue of their isolation are now gone'[40] (illus. 197, 198 and 199). The same Reuters report cited the coral reef biologist John W. McManus (who two decades earlier had foreseen claims of an EEZ in the area[41]) about China's 'reclamation' activities as constituting 'the most rapid rate of permanent loss of coral reef area in human history', referring to the 13 sq. km (5 sq. mi.) covered since 2013, much of it now overlaid by concrete. McManus himself poignantly quoted Joni Mitchell's 1970 lyric regarding the paving of paradise in favour

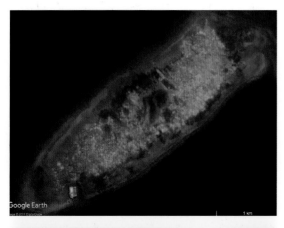

humans pursuing geopolitical ends who 'out of water brought forth solid rock', with yet-unmeasured consequences for local reefs.

The Malé International Airport in the Maldives archipelago north of Diego Garcia has a 3,200-m (10,500-ft) runway underlaid with coral mined from the local reef flat. Elsewhere, coral slabs are still quarried, ground up and used as an aggregate with bitumen in repairing and restoring remote Pacific airfield runways. To the airport construction engineer, coral remains a mineral – 'an organic sedimentary rock' – an incidental and locally sourced product of a living organism.

Enlarging Malé's airport was prompted by the booming of tourism in the Maldives, and corals, formerly used to build mosques and memorials in cemeteries, saw increased use in constructing luxury tourist accommodations, 'which typically consist of one hotel on a previously uninhabited island'.[42] The private islands often are surrounded by a breakwater of coral rock mined from the living reef. At least one high-end resort offered guests (without apparent irony) the opportunity to adopt and name a rescued coral fragment from an underwater nursery, for eventual placement on a restored reef. There and in the Caribbean, fashionable hotels at reef sites still are constructed of ancient coral, or incorporate modern coral rock as aesthetic accents.

Coral Gables, Florida (part of greater Miami), is nothing if not fashionable. Planned in the early 1920s as an elite leisure community, many of its buildings were constructed of oolite, a local limestone composed of conglomerated grains of degraded coral rock, shells and other calcareous materials. Erroneously assumed to be the foundation of an ancient coral reef, the stone gave its misnomer to the community. Downtown, many buildings were given a more high-end appearance by facing them with an expensive and striking stone of true fossilized

of parking lots. In this case, it was not the coral polyps, working as in 'The Pelican Island', James Montgomery's 1828 poem discussed earlier, to build a memorial to themselves, but rather

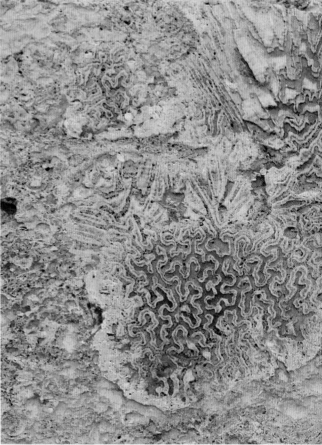

coral, Key Largo limestone or Keystone, quarried in the Florida Keys. The combined fire and police station built at the end of the Great Depression made extensive use of this stone (illus. 200 and 201). Looking at the building's blocks of fossilized reef coral that grew at the last high stand of sea level 130,000 years ago, coral reef ecologists and geologists can read records of coral growth, bioerosion and storm disturbance events in the ancient environment.[43]

Further south on U.S. Route 1, towards the Florida Keys where the ornamental Keystone was quarried, lies a tourist-attracting complex of carvings of the less voguish, unadorned local oolite. Beginning in 1923 and continuing for nearly three decades, Edward Leedskalnin, a diminutive, lovelorn Latvian immigrant, built a complex of structures carved from about 1,000 tons of this material.[44] Individual pieces weighed from several to some 27 tonnes (30 short tons). Inside the wall of megaliths that surrounds the Coral Castle are a 12-m (40-ft) obelisk, crescent moons and planets, and a 'telescope' (illus. 202). More domestic objects such as rocking chairs, a bathtub, a bedroom and a heart-shaped table lend credibility to the builder's claim that he undertook the huge project in memory of his unrequited love of a young woman, whom Billy Idol recalled in his song 'Sweet Sixteen'. The music video opened with a well-known photograph of Leedskalnin standing amid his creation, with the added epigraph 'Love turned to Stone'.

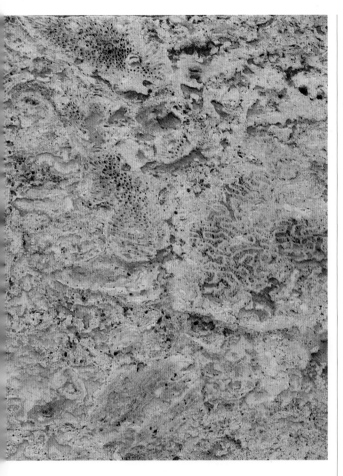

Far left: 200 The tower of the Coral Gables public safety building in Florida (now the Coral Gables Museum), built of Key Largo limestone and opened in 1938. The exterior of the museum provides an opportunity for geology students to peer into the interior of a coral reef. Left: 201 Detail of a block of limestone used in the construction showing fossilized coral colonies cut in several orientations. Each block tells a story of growth, erosion, sedimentation and cementation.

Beyond an eccentric's oolitic folly, 'fanciful' certainly applies to the surprising use of the coral motif at the World's Fair of 1939 in New York City, which offered glimpses of 'tomorrow's world – today!', including such firsts as television, fluorescent lights, air conditioning and nylon. Separate from the fair's government and corporate zones was a less lofty 'Amusement Area' that included carnival rides and girlie shows, and a rare, ephemeral example of Surrealist architecture.

Lewis Kachur tells the story in *Displaying the Marvelous: Marcel Duchamp, Salvador Dalí, and Surrealist Exhibition Installations* (2001). New York gallery owner Julien Levy, together with Salvador Dalí (whom André Breton had already anagrammatically nicknamed Avida Dollars) and other investors, formed a corporation to design and construct a pavilion suggestively called Bottoms of the Sea, which would include large aquariums where topless, scantily clad mermaids and sirens – Lady Godivers – frolicked.

Dalí's design for the pavilion included a facade featuring a huge black-and-white version of Botticelli's Venus (alas, not with the head of a fish as the inverted mermaid that Dalí intended), and the pavilion was renamed the *Dream of Venus*. The striking facade included a fringe of huge stuccoed branches of coral. Night-time photos of the illuminated pavilion showed stark forms and ominous shadows (some seeming like grasping hands) cast by the corals.

Others of Dalí's plans were not realized or were changed after he returned to Europe. But the coral remained, and even metamorphosed and proliferated: in the niche where Venus had been there later grew a huge branch of red coral supporting larger-than-life images of a mermaid and a bathing beauty (illus. 203), and sometimes populated by living touts in swimming attire to attract an audience. It was this later version that was shown, snow-covered in the dead of winter, on the cover of the *New Yorker* magazine for 24 February 1940.

For Dalí, coral meant more than a motif in keeping with the pavilion's marine theme,

Opposite page (top): 202 Coral Castle in Leisure City, Florida; (bottom) 203 Detail of the later facade of Salvador Dalí's 'Dream of Venus' pavilion at the 1939 World's Fair.

whatever its title. He was interested in organic architecture, and growing coral was an obvious and meaningful construct. The art critic David Cohen described it as an 'organic facade that was a gaudy riff on Gaudí' (Dalí's Catalan countryman who incorporated marine forms and other biomorphic elements into his visionary architecture). But it seems to have gone deeper; after returning to Spain from Paris to recuperate from an illness, Dalí wrote in his autobiography, using coral imagery that by now is familiar:

I actually felt a kind of transparency, as though I could see and hear all the delightful little viscous mechanisms of my reflowering physiology. I had the illusion of having an exact consciousness of the circulation of my hard blood through the tender and ramified tubes which I felt covering the euphoric curve of each of my shoulders, like epaulettes of living and subcutaneous coral imbedded in my flesh.[45]

The intellectual adoption of corals by the Surrealists – as a crystal by André Breton and as part of his own body by Salvador Dalí – came when coral was metamorphosing again in the imagination, in what would become a New Age of Corals.

7 A New Age of Corals

It's past time to tell the truth about the state of the world's coral reefs... They have become zombie ecosystems, neither dead nor truly alive in any functional sense, and on a trajectory to collapse within a human generation.
Roger Bradbury, 'A World Without Coral Reefs', *New York Times*, 13 July 2012

An Eternal Reef is a designed reef made of environmentally-safe cast concrete [and] combines a cremation urn, ash scattering, and burial at sea into one meaningful, permanent environmental tribute to life... It seems more of a *beginning* than an ending.
Eternal Reefs, Inc., www.eternalreefs.com

Evolutionarily distinct coral lineages have been viewed and treated separately for millennia – red coral as a precious stone, talisman and healing substance, and scleractinian hard corals as builders of reefs that threatened seafarers and often entombed them, but also produced new land where there was none. Their discovery by seagoing explorers and later by adventurers and tourists transformed coral reefs into idyllic oases of abundant, brilliant, charismatic and diverse life forms that were the evidence of riotous evolution or the gift from a provident creator. Now, in the Anthropocene epoch when the human population exceeds seven billion and its ineluctable influence imperils the biosphere, precious, deep-sea and tropical reef corals have converged in a new age as the epitome of what is being lost to the realization of such hazards as economic overexploitation, pervasive pollution and a changing climate (the greatest existential threat to corals).

Reef-building Corals in a New Age

The half-century between Cook's navigation of the Labyrinth and Darwin's 'ah-ha' experience aboard the *Beagle* was not only historian Richard Holmes's 'Age of Wonder' but also an Age of Corals when their nature and global geography emerged. Two centuries later we are in a New Age of Corals when deep concerns over coral reefs are surfacing. Some areas of the Indo-Pacific, where most of the world's coral reefs are located, lost between 1 and 2 per cent *per year* of coral cover (the percentage of a surveyed area of seabed that living corals occupy) between 1968 and 2004; many Caribbean reefs saw total declines of 80 per cent from 1977 to 2001, and even on the well-managed Great Barrier Reef (GBR), coral cover decreased by 51 per cent between 1985 and 2012.[1] About three-quarters of the world's coral reefs are at medium or high risk of further degradation, according to the World Resources Institute.[2] In its report 'Climate Change 2014', the Intergovernmental Panel on Climate Change (IPCC) deemed coral reefs among the world's most threatened ecosystems. The peril is greatest in Southeast Asia, where nearly 30 per cent of the world's reef area occurs, and almost 95 per cent of those reefs are at risk. In Indonesia nearly a quarter of the population of 250 million relies on coral reefs for its food and livelihood.

Coral's reciprocally accommodating, symbiotic nature was discovered at a time of pervasive industrialization and concomitant

Opposite page: 204 David Stacey, Living in Paradise, *c. 1988, acrylic. Overleaf: 205 Zane Saunders,* Coral Serpent, *1990, linocut print on paper.*

3/10 "Coral Serpent"

Jane Saunders 1990

social change, including a focus on city planning as urban populations grew. In his essay 'How to Prosper in a World of Limited Resources' (1998), Eugene Odum used the mutualistic coral symbiosis as a moral lesson. He also pointed to tropical rainforests, like coral reefs exemplars of luxuriant 'hotspots' of biodiversity and beauty that succeed through recycling. Resplendent coral reefs in their New Age became, first, visions of harmonious nature and sustainability, emblematic of Green environmental awareness; then a prospective fount of 'drugs from the sea'; and later both the 'canary in the coal mine' and the poster child for nature at risk in a world gone environmentally awry.

Coral reefs-as-rainforests saw explicit visual expression in *Living in Paradise* by David H. Stacey, a Queensland artist familiar with the Great Barrier Reef just offshore from the Daintree Rainforest, near Cook's Endeavour Reef (illus. 204). It was a small step from icon to caricature, however, and comforting coral reefs in tropical touristic art and contemporary culture became populated by ever more winsome fishes, sea turtles and mermaids (recall the Disney characters in *Finding Nemo/Dory* and *The Little Mermaid*), as well as dolphins with their rictus of bonhomie.

Corals are rarely depicted in traditional Aboriginal Australian art but are appearing increasingly in modern printed media during a new age of reef environmentalism, tourism and aesthetic expression. Even the Rainbow Serpent, a powerful ancestral being traditionally associated with freshwater sites, had a marine origin[3] and a contemporary marine incarnation, both in times of rapid climate change and sea level rise. A remarkable series of images, *Coral Serpent* (1990) by Zane Saunders, who grew up in North Queensland, presents a literal and vibrant reef scene with diverse fishes, corals and other invertebrates in which the Rainbow Serpent spans them all (illus. 205).

According to Saunders, his early paintings of the reef reflected his experience with that magical environment as a young Aboriginal Australian person, fishing there and exploring its natural wonders and those of the nearby rainforest. He and many Indigenous Australians (for whom the Rainbow Serpent is a unifying symbol) retain the deep concerns that followed colonization – the loss of freedom and of Aboriginal law and governance of themselves and the land, and the balance of the natural environment. It was as though 'the Rainbow Serpent was speaking with my spirit, inspiring me . . . [to show] the GBR that meanders and follows the shore line off the east coast of Australia, like that of a serpent resting but alive, giving life, generating a Dreaming that is continuous today, unending.'[4]

Artistic depiction of corals has become aimed at raising awareness in a time when much of society has lost its connection to nature and also lacks the scientific literacy needed to appreciate the drivers and impacts of global climate change. The American artist Alexis Rockman's meticulous, sometimes apocalyptic paintings melding art and science are informed by evolutionary and environmental themes, often having what Stephen Jay Gould called an 'explicit acknowledgment of inevitable human presence'.[5] In *The Pelican* (2006), Rockman combines the hazy, stifling Anthropocene vision of coastal industrial ruins above the murky remains of a reef ravaged by debris and toxic waste (illus. 206).

The ceramic sculptor Courtney Mattison calls herself an 'artivist', working at the interface of coral reef conservation and popular culture. For Mattison and a growing (but still minuscule) number of artivists versed both in art and science, art can move people emotionally and inform them in ways that scientific data cannot: 'We won't act unless we care, and we won't care if we don't know.' Her massive, detailed ceramic depictions

206 Alexis Rockman, The Pelican, *2006, oil on wood.*

of healthy coral reefs transitioning to degraded states (and in places recovering) bear witness to the plight of reefs worldwide, not as 'memorials to reefs for a future without them, but rather

207 *Courtney Mattison,* Our Changing Seas II, *2014, glazed stoneware and porcelain.*

celebrations of their enduring yet fragile health and beauty' (illus. 207).[6]

The 'Crochet Coral Reef' was conceived and curated by Queenslanders Margaret and Christine Wertheim to draw attention to the modern-day decline of reefs, especially their beloved Great Barrier Reef. The collaborative endeavour combines aesthetics and activism, and involves thousands of contributors worldwide who, like individual polyps each adding its 'gram or two of lime', crochet coral polyps and colonies, sea anemones, sea slugs and other organisms to create an archipelago of reefs and atolls that has been growing since 'spawned' from the original in 2005.[7]

As a natural product growing in idyllic settings and associated with a healthy outdoor lifestyle, reef corals eventually entered the age of new medicine. The appearance of reef corals as a dietary supplement is more recent than the historical medicinal uses of red corals and is fraught with controversy because of zealous, sometimes fraudulent, hyper-marketing of 'coral calcium' in the heyday of coral reefs as the image of hale well-being. The most notorious case was that of a convicted serial scammer (now under a ten-year jail sentence with '10 easy payments of one year each'[8]) whose televised 'infomercials' outrageously claimed that 'Coral Calcium Supreme' was a cure for 'cancer, heart disease, multiple sclerosis, lupus, and other serious ailments'.[9]

An expert guest interviewed on some of the infomercials expounded extensively on the supposedly salubrious properties of 'coral calcium' and the basis for its effectiveness. The attributes included promoting health and longevity, purportedly evinced by the absence of cancer and degenerative diseases in the healthy, long-lived natives of Okinawa, Japan, who have high dietary intake of calcium where much 'coral calcium' is mined ('ecologically', of course – no living corals are harmed in producing the powder and pills). Trace minerals (not always prominently listed – one of them was lead) in the 'elixir of the new millennium' have been touted for their presumably healthy effects, rarely with any specifics.

In a reversal of Shakespeare's verse, in the new age of biomedicine 'of corals are bones made.' This transformation refers not just to dietary 'coral calcium' to replenish bone minerals, but to using scleractinian skeletons as substitutes for bone grafts in humans, a practice dating back more than three decades. Coral skeleton was conceived as an alternative transplant material to avoid potential immunological responses and transfer of human disease. The skeletal structure of *Porites*, for example, is similar to that of human bone, with which it is biocompatible. Moreover, its porosity allows the invasion of bone-producing cells and growth factors from the recipient, eventually supplanting the coral with new bone as the implant is biodegraded and rebuilt. Deep-sea bamboo corals also are promising as implants, prompting the suggestion that they be grown in marine bone farms.

Replacement eyes made of artificial materials, including glass, could not be attached to the extraocular muscles and thus could not move in the eye socket, detracting from a natural look. Moreover, the devices were not permeated by the recipient's blood vessels and eventually were rejected by the body. A chemically modified coral skeleton where the carbonate moiety is replaced by the phosphate of the human bone mineral hydroxyapatite avoids these problems and is used to fashion the spherical Bio-Eye® orbital implant. Once integrated with the body, the moveable, less degradable, bone-like implant can have a realistic artificial eye attached to its anterior (illus. 208).

Thus, in the new, once-future, medicine, corals for the anthropologist Stefan Helmreich have become tissue donors to humans.[10] What would Dalí have made of this?

208 The Bio-Eye® orbital implant of modified coral skeleton having the same microarchitecture as human bone, with a rendering of the visible anterior portion of an artificial eye that would be attached to the implant.

Atolls Awash

The image of intermingled human bones and corals is pervasive. In Miami, Florida (average elevation 1.8 m/6 ft), where parts of the city are regularly flooded as sea level rises at ten times the global average rate, the film-maker Lucas Leyva combined these themes in a video produced in

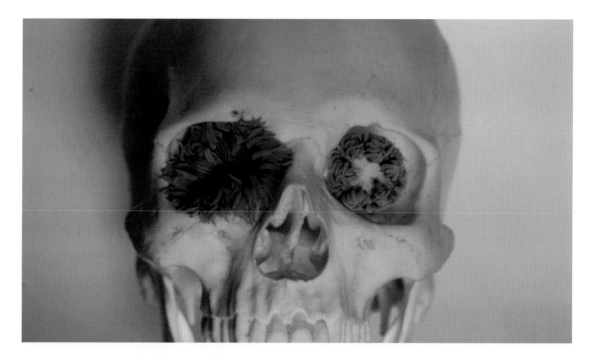

2014 with the Miami science and art collaborative Coral Morphologic, *The Coral Reef Are Dreaming Again (Zhuangzi Dreams of Coral Reefs)* (illus. 209). 'It takes place hundreds of years in the future, after Miami is completely under water. It's about corals living on our bones, the same way as we're living on their bones now.'[11] According to the title of John McSwain's 2014 article for *Vice* magazine, 'Miami is Drowning, and the Corals Couldn't be Happier.'

Miami (built on a geological foundation of porous oolite) is not the only coastal city that is drowning as sea level rises – the result of both the thermal expansion of seawater, which occupies a larger volume as it warms, and the melting of glaciers and the polar ice caps, which adds water to the oceans. Since the early 1980s the focal issue in global climate change has been the rise of atmospheric and sea surface temperature (global warming), driven by the burning of fossil fuels that adds the principal greenhouse gas, carbon dioxide (CO_2), to the atmosphere – a staggering 2 trillion metric tons since the advent

209 *Still from the short film* The Coral Reef Are Dreaming Again *(dir. Lucas Leyva, 2014). The eye sockets are inhabited by 'flower anemones' (family Phymanthidae), anthozoan relatives of reef-building corals.*

of industrialization in the mid-eighteenth century. The added CO_2 increasingly traps solar heat rather than allowing it to radiate back into space. Most of the heat ends up in the oceans – 93 per cent of the additional heat since the 1970s.[12] This oceanic warming is leading to rising sea levels, mass bleaching of corals and higher incidences of coral diseases.

Absent technological and engineering solutions, Miami (already installing pumps and raising its streets and pavements with their lovely coral kerbstones a step ahead of sea level) and other low-lying cities may have to depopulate and move people inland as their infrastructure is swamped.[13] But relocating to adjacent higher ground is not an option for citizens of the Maldives, a coral atoll nation in the Indian Ocean comprising more than

1,100 islets having an average elevation of only 1.5 m (just under 5 ft), or the Pacific nations of Tuvalu and nearby Kiribati, the latter including 33 atolls and flat islands scattered across 3.5 million sq. km (1.35 m sq. mi.) of ocean. The persistence of these coral-built foundations, which cannot rise above mean local sea level because corals tolerate only brief exposure to air, depends on their rates of calcium carbonate production exceeding those of reef erosion. And if sea level rises, the structures already built by humans on these foundations will be immersed.

In 2009, then president Mohamed Nasheed of the Maldives (where *atholhu* in the local Dhivehi language gives us the English word atoll) held a meeting of his cabinet underwater, in a coral lagoon, where he signed a document calling for cuts in CO_2 emissions. The message was that without such cuts to slow the pace of climate warming and the resultant rising sea level, inundation would be the fate of the Maldives and other Indo-Pacific coral archipelagos by the end of the twenty-first century (illus. 210).

Nasheed was pragmatic in confronting projected sea level rises and investigated buying land in other countries as a new home for the Maldivian population (in 2014, an estimated 394,000) when their nation is awash. Facing similarly dire prospects of inundation, Kiribati's then president Anote Tong in 2014 purchased 22 sq. km (8.5 sq. mi.) of high ground in Fiji, seeing 'migration with dignity' for the population of 110,000 as more cost-effective than building barrier defences for the many islands of his far-flung nation.[14] (illus. 211) The Kuna community of the San Blas Islands will relocate to the foothills of their Caribbean coastal strip on the Panamanian mainland when their islands become uninhabitable in coming decades.

Below: 210 The Maldives' capital, Malé, where 120,000 people live about 1 m (3 ft 4 in.) above sea level; Opposite page: 211 A man on South Tarawa in Kiribati rebuilds the sea wall of coral rock that protects his home from the rising sea level.

The UN Climate Change Conference in Paris (formally, the 21st Conference of the Parties to the United Nations Framework Convention on Climate Change, or COP21) was held in Paris in December 2015. Even the optimistic forecasts offered in Paris, based on the intended nationally determined contributions to reducing carbon emissions, foresee temperature increasing on average by 2.7°C and as much as 3.5°C,[15] and sea level rising 2.4 m (nearly 8 ft), by the end of the twenty-first century. These changes assume that the Paris Agreement, ratified and now in effect, will be extended and strengthened beyond 2030.[16] This temperature increase would exceed the 1.5°C limit promoted hopefully by the coalition of the Alliance of Small Island States and the Least Developed Countries, and even the wider political target of a 2°C limit adopted at COP21. These projections may spell doom for many inhabited atolls in the Indian and Western Pacific Oceans, where the water's rise will be greater than average. Even before atolls go under, they will be eroded by rising seawater and El Niño-driven superstorms, and many will be abandoned because of their limited amount of tillable soil and fresh water.

And what about coral reefs themselves? How they fare under rising sea level will depend on whether their upward accretion can match the pace of inundation. Their rapid growth, and characteristic fragmentation and regeneration of branches to form new colonies, makes acroporid (staghorn) corals dominant reef builders, well suited to keeping up with the projected rapid rise in sea level, as they have done especially in the past 1.8 million years.[17] But these corals also are especially susceptible to worsening human disturbances and are on the decline worldwide, thus becoming less able to help keep reefs from drowning. As it is, more than half of coral reefs worldwide are falling behind the rate of sea level rise. Ocean warming and acidification are projected not only to reduce the growth rate of most corals but also to leave them producing weaker, more porous skeletons, which will render reefs more susceptible to bioerosion by boring organisms and damage from ever-stronger storms. Ultimately this enfeeblement will compromise the coral ramparts that protect coasts against storm flooding and shoreline erosion.

Shifting Baselines, the Slippery Slope to Slime, and Debt Restructuring

Divers in the Maldives and elsewhere never forget their first encounter with spellbinding coral reefs, and it is this memory that for many sharpens the urgency of protecting these 'pristine' ecosystems. But there is a problem, because

> everyone, scientists included, believes that the way things were when they first saw them is natural. However, modern reef ecology only began in the Caribbean, for example, in the late 1950s . . . when enormous changes in coral reef ecosystems had already occurred. The same problem now extends on an even greater scale to the SCUBA diving public, with a whole new generation of sport divers who have never seen a 'healthy' reef, even by the standards of the 1960s. Thus there is no public perception of the magnitude of our loss.[18]

This reality is the 'shifting baseline syndrome', expressed as long ago as 1879 by the poet Gerard Manley Hopkins as 'After-comers cannot guess the beauty been.' Setting today's baseline all too often ignores yesterday's. Then U.S. president Barack Obama unintentionally provided an example in his climate interview in *Rolling Stone* of 8 October 2015, when he recalled snorkelling as a child in Hawaii on 'coral reefs that . . . were lush and full of fish that now . . . are not'. But reefs around an urbanized Oahu were *already* deteriorating by the 1960s–70s baseline of his boyhood.

A shifting baseline or reference point makes it hard to judge what factors are involved in coral reef decline today, because many local and regional impacts, such as overfishing, pollution and agriculture, pre-date global warming and ocean acidification by decades to millennia.[19] There are few, if any, truly pristine reefs left in the world to serve as a baseline by which to gauge what already has been lost in the Anthropocene, and to guide management to sustain and restore reefs. Remote, uninhabited coral islands may provide a glimpse of what healthy reefs looked like in the past. Non-calcifying fleshy seaweeds and turf algae characteristically dominate inhabited islands in five central Pacific archipelagos, whereas uninhabited islands have more active reef builders, such as corals and crustose coralline algae. Seemingly it is impacts of the human inhabitants that are causing the difference.[20]

My own experience with the Great Barrier Reef coincides almost exactly with the 27-year decline in coral cover by 51 per cent from 1985 to 2012. Although mass coral bleaching and destructive tropical cyclones were involved, so was the crown-of-thorns sea star, *Acanthaster planci*, a nightmarish predator that everts its stomach over corals and digests their tissues and which was seen as the major threat to the reef during outbreaks in the 1960s and beyond. On my first-ever dive there, the most striking sight as I neared the bottom was a front of these sea stars advancing across healthy plating acroporid corals, leaving behind digestively denuded, deathly white skeletons (illus. 212). A bone of contention has been whether outbreaks of the crown-of-thorns (for millennia a denizen of the reef) were part of a natural cycle or whether they were caused by agricultural runoff that enhanced growth of the planktonic algae eaten by larval sea stars.[21] People having long experience with these reefs confirm that they now are not what they were, and that the crown-of-thorns has been only part of the problem. This trend is evident even to a diver and author having only a quarter-century's personal perspective on the GBR.

Less debatable is the role of fishing in coral reef decline. Severe overfishing that includes trapping or spearing and removing herbivores such as parrotfishes often leads to overgrowth of corals by fleshy seaweeds, particularly in areas where poor land use practices increase runoff, enhancing algal growth (illus. 213). Depleting Caribbean reefs of their fishes allowed large populations of algal grazers such as the sea urchin *Diadema antillarum*. When most of these urchins died in an outbreak of disease in the 1980s, algal overgrowth increased. Eventually the reef-building corals were smothered and replaced by seaweeds in what ecologists call a 'regime shift'. In extreme cases, reefs were degraded to bacterial and algal slime-covered rubble. Avoiding this 'slippery slope to slime' has been a mantra for environmentalists concerned for coral reefs in the Anthropocene.

Recognizing that international commercial fishing, mining and drilling for oil increasingly threaten coral habitats and broader marine ecosystems, nations have begun to shelter them as preserves and marine protected areas (MPAs), hoping to slow or reverse the baseline shifts. Among Obama's last official actions in office was to create the Northeast Canyons and Seamounts National Monument off Cape Cod in September 2016. More than seventy species of azooxanthellate cold-water corals (some not known from anywhere else) have been identified in these 12,725 sq. km (4,900 sq. mi.) of seabed, where oil drilling, mining and some commercial fishing will be prohibited.[22] The hope for such protected areas is that marine species, including corals, will be better able to contend with climate change when freed from these additional stresses and that the protection will help fish stocks to recover and spread.

Tropical island nations such as the Seychelles, northeast of Madagascar in the Indian Ocean, with its 115 islands and many coral reefs, are particularly dependent on marine economic

212 Crown-of-thorns sea stars devouring the polyps of a plate-like acroporid coral on the Great Barrier Reef, 1983.

Opposite page: 213 Poaching of fishes (especially parrotfishes, as well as snappers, a triggerfish and a moray eel) and invertebrates (lobsters) contributes to the decline of Caribbean coral reefs (here in a national marine park in the Dominican Republic) down a slippery slope to slime. Divers working from one boat may take 10,000 lb (4.5 metric tons) of such fish per week.

activities both for domestic consumption and often to pay their international debts. With the cooperation of the Paris Club of major creditor countries, and with grants and a concessionary loan from The Nature Conservancy, the Seychelles established a conservation and climate trust that bought back U.S.$21.6 million of its sovereign debt.[23] Some of the country's long-term, lower-rate payments to the trust will fund conservation and 'climate adaptation' projects; other interest payments will be invested in a trust fund for future conservation projects once the debt is paid off. The restructuring hinged on a 'debt-for-nature' deal whereby 30 per cent of the country's exclusive economic zone (EEZ) of 1,374,000 sq. km (530,500 sq. mi.) will be designated as MPAs for conserving biodiversity, where sustainable fishing and other marine activities (infrastructure, renewable and non-renewable energy, and tourism) will be managed and regulated.

Coral Bleaching and Carbon Bargaining

A wider consequence of continued warming despite the start of controls of greenhouse gas emissions will be an increasing frequency of mass coral bleaching, a driver of reef decline and demise. As we have seen, the bleaching of corals results from the degradation or loss of their endosymbiotic algae under stressful conditions. Loss of the deeply pigmented zooxanthellae leaves behind the translucent coral animal tissues, through which the bone-white skeleton becomes visible, giving the condition its name (illus. 214). The stressors involved are myriad, acting individually, additively or synergistically, and beyond extreme temperature, irradiance and salinity include sedimentation by runoff from agricultural or clear-cut lands, bacterial infection and diverse forms of pollution.[24] Runoff and pollution are the greatest immediate threat to the resilience of the GBR under manifold stresses.

Photooxidative damage by bright sunlight exacerbates the effect of elevated temperature, and the two obviously are correlated because temperature and solar irradiance are both highest in summer. This is why corals in shallow, well-lit areas may bleach while those nearby in deeper water experiencing less light do not, even though the water temperature is the same. Because corals cannot avoid warmer water in the decades ahead, they may become restricted to slightly greater depths where light levels are lower, to avoid the one-two punch of these stressors.

Not all corals are equally susceptible to bleaching: branching species bleach more readily than massive and encrusting species,[25] leading to changes in coral community composition. Nor is bleaching all-or-none: a given coral colony may lose some, but not all, of its algae and recover if the stress is not prolonged. But when thermal bleaching is severe and protracted, if the coral does not regain algal symbionts (its principal source of nutrition), it will starve, weaken and die. Bleached corals also are more susceptible to disease, and their growth and reproduction suffer even if the corals survive.

Accounts of isolated bleaching date back at least to the 1928–9 Great Barrier Reef Expedition to the Low Isles, but the instances invariably were confined events traceable to a particular local stressor. In the early 1980s coral bleaching occurred on much larger geographic and temporal scales, beginning with those cases associated with El Niño warming events in the eastern Pacific off Panama, first documented and then demonstrated

214 Coral bleaching in New Caledonia in March 2016.

experimentally by Peter W. Glynn.[26] Reports of other 'mass coral-reef bleaching' events began to increase in frequency.[27] In an event throughout the tropics in 1997–8, bleaching killed 16 per cent of shallow-water coral reefs worldwide, and more than 90 per cent of corals in shallow water in the most severely impacted areas of the Indian Ocean. A second global bleaching event occurred in 2010, another visually striking effect likely attributable to global climate change.

The initial syntheses further implicated high-temperature anomalies, because mass bleaching occurred in ocean 'hotspots' having a sea-surface temperature (SST) more than 1°C above long-term local averages during the warmest months.[28] Corals worldwide that collectively spanned a maximum summer temperature range of 9°C were each found to be living within 1–2°C of their lethal limits (illus. 215).[29] The thermal sensitivity of corals and forecasts of warming seas ahead indicate 'that bleaching could become an annual or biannual event for the vast majority of the world's coral reefs in the next 30–50 years'.[30]

Understanding the association of coral bleaching with local seawater temperature was refined by analysing data for hundreds of reefs along the 2,300-km (1,400-mi.) length of the Great Barrier Reef, including those that bleached during events in 1998, 2002 and 2016. The different geographic 'footprint' in each of these events is explained by the magnitude and spatial coverage of SST anomalies. Repetitive sampling of 171 individual reefs in all three episodes reveals that the degree of bleaching on a particular reef is due to the local severity of the temperature anomaly.[31]

The zooxanthellae living inside the host coral's cells are the weak link in the chain of events leading to coral bleaching. At high temperature in bright light, photosynthesis in the algae becomes inhibited, and energetically excited electrons instead produce toxic reactive oxygen species (ROS) that overwhelm the algal antioxidant defences. ROS subsequently leak from the algae to the animal host, adding to its own burden of oxidative stress, damaging or killing

its cells and leading to loss of the algae from the coral.[32]

Scleractinian corals worldwide collectively host diverse clades (genetic lineages) of symbiotic dinoflagellates (mostly *Symbiodinium*) that differ in their susceptibility to photoinhibition at elevated temperatures. Field studies correlated the distribution of the zooxanthellae (in the local reefscape and on geographic scales) with their tolerance for high temperature and irradiance.

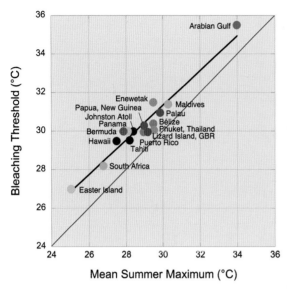

215 The relationship between the mean summer maximum temperature of seawater and the bleaching threshold in reef corals from different regions. Various colours and shades designating the individual sites denote different regions of several oceans (blues = Pacific Ocean; greens = Indian Ocean and Arabian/Persian Gulf; purples = Atlantic Ocean and Caribbean Sea). The diagonal red line is the hypothetical relationship where bleaching threshold equals mean summer maximum temperature. The black line is the best-fit regression for all sixteen sites; the position of the data points and the regression line above the red line indicates that collectively the coral bleaching threshold is about 1–2°C higher than the local mean summer maximum temperature. Opposite page: 216 Paired photographs before and after the 2015 mass bleaching event in Guam.

A single coral colony of thousands of polyps might simultaneously host one or several algal clades, the predominance of which could change following a bleaching event in what has come to be called 'symbiont shuffling'. Thermally resistant clades come to dominate in some cases, leading to the identification of algal lineages thought to be genetically 'adapted' to high temperature. Colonies in which heat-tolerant zooxanthellae had become predominant did not bleach in subsequent episodes when other colonies did. The physiological flexibility afforded by such shuffling would not protect corals against temperature increases projected in the long term, nor is it known how many species of corals can host the resistant algae, but for some it offers 'a nugget of hope' for coral reefs by buying time while there is still a chance to put controls on the emission of greenhouse gases.

Corals may be running out of time to buy. When the beans were counted at COP21, it emerged that the pledges by 196 nations or parties (including the multinational EU) to reduce carbon emissions would yield cuts only about half as large as those needed to hold warming to the agreed target of 2°C by 2100, with a likely actual increase of 2.7°C if most nations strengthen and extend their pledges, or greater if they do not.[33]

In 2014, the National Oceanic and Atmospheric Association (NOAA) had listed a further twenty coral species, bringing the total to 22 considered as 'threatened', for protection under the U.S. Endangered Species Act. Soon afterwards it was widely reported that seawater temperatures in 2014, again in 2015 and yet again in 2016 were the warmest on record, and that mass coral bleaching had begun in the Pacific (including first-time bleaching in the Hawaiian Islands, followed by an exceptional second event there in 2015). The forecast by NOAA's Mark Eakin for globally widespread bleaching in 2015–16 was realized when bleaching continued to spread in October 2015 and in early 2016 became the longest recorded global bleaching event.[34] (illus. 216) Following nearly three months of seawater temperatures a degree above the normal summer high, in March 2016 the Catlin Seaview Survey photographed widespread coral bleaching off Cape York at the remote northern end of the Great Barrier Reef, where formerly exemplary

reefs were 90–100 per cent bleached,[35] and two-thirds of the corals along a 700-km (435-mi.) stretch eventually died. On Kiribati's equatorial Christmas Island, record warming 3°C above normal in the 2015–16 El Niño killed 85 per cent of the island's corals.[36] Once this third global bleaching event ended by June 2017, some reefs had bleached in three consecutive years.

Australia, despite its national passion for the Great Barrier Reef, has one of the highest per-capita carbon emissions and dragged its feet on reducing them (a conservative coalition even scuttling an earlier carbon tax), so that other nations pointed fingers and questioned the commitment of prime minister Tony Abbott's government to combating climate change. In addition to over-fertilization by agricultural runoff that promotes the growth of fouling algae, the declining health of the GBR is related to the rapid industrialization of the Queensland coastline. This development involves increasing dredging and construction of ports for the maritime export of coal (some of the world's largest reserves being in Queensland's Galilee Basin) and liquefied natural gas, on which Australia relies heavily both for trade dollars and domestic energy.

Widespread outcry caused the Queensland government to block the planned dumping of dredge spoils in the GBR World Heritage Area (WHA), but UNESCO's World Heritage Committee remained sufficiently concerned that it threatened to place the WHA on its 'in danger' list. This threat spurred the national and Queensland governments to prepare the Reef 2050 Long-term Sustainability Plan, the final version of which was released in March 2015 following input from the public and, notably, the Australian Academy of Science, which had deemed the draft report's proposals inadequate to restore or even maintain the diminished GBR. The revised plan repeatedly acknowledges the dangers of climate change but offers no actions in that regard. Environmental groups assert that the plan does not even go far enough towards improving water quality and is seriously underfunded.

Through unprecedented diplomatic lobbying, the Australian government successfully avoided the GBR being deemed 'in danger' in the June 2015 report by UNESCO. However, the GBR remains on UNESCO's watch list, and the organization did require the Australian government to report on progress by December 2016, with the possibility that a stigmatizing 'in danger' listing would be reconsidered in the absence of adequate progress. It was against this backdrop that Tony Abbott was ousted as Australian Liberal Party leader and ultimately resigned as prime minister.

His successor Malcolm Turnbull's government largely held to the earlier emissions-reduction programmes and their modest target. With a majority of Australians coming to favour a price on carbon emissions, the Labor Party opposition, in the aftermath of Paris and with elections upcoming in 2016, pledged greater reductions and to be carbon-neutral by 2050, and promised substantial new funding to remediate the GBR.[37] Regardless, Turnbull and the Liberals were victorious in the June 2016 election; the low emissions-reduction target remains, as does the spectre of opening new coal mines in the Galilee Basin.

Despite UNESCO's watchful eye, the Queensland government soon thereafter declined to protect Cape York from further land clearing that would have spared the GBR from additional runoff, a failure that became a red flag in the December 2016 progress report. Conceding this negligence, Queensland environment minister Steven Miles could only hope that his government's recommitting to passing tougher tree-cutting laws would satisfy UNESCO, but he realized that this passage depended on winning a majority in the next parliamentary election.[38]

Meanwhile, in the U.S., the November 2016 presidential election gave cause to worry about the new administration's commitment to the global effort towards mitigating climate change. On record as seeing climate change concerns as a foreign hoax to disadvantage American industry, president-elect Donald Trump warned that he might ignore the Paris Agreement and nominated climate-change 'sceptics' and cheerleaders for the fossil-fuel industry to head the Environmental Protection Agency (EPA) and the Department of Energy, and the chairman of ExxonMobil as his Secretary of State and chief U.S. climate emissary. At their confirmation hearings in January 2017 all grudgingly admitted the reality of human involvement in global climate change, and although 'none seemed eager to seek a solution,'[39] the Senate confirmed all of them. Moments after the new president was inaugurated, the White House website had displayed a promise to scuttle the Obama administration's Climate Action Plan and to refocus the EPA in an 'America First' energy plan that did not mention renewable sources. Scarcely two months later, the president signed orders to begin to fulfil that promise. At a press conference on 16 March 2017, the director of the White House Office of Management and Budget summarized the view from the top: 'As to climate change, I think the president was fairly straightforward – we're not spending money on that anymore. We consider that to be a waste of your money to go out and do that.' In August 2017 President Trump announced that he was withdrawing the U.S. from the Paris Agreement and seeking a better deal.

'The Other Carbon Dioxide Problem'

During the past two decades of escalating atmospheric CO_2 levels, an additional concern has gained attention: about one-third of the

CO_2 added to the atmosphere has dissolved in the oceans. CO_2 leaving the atmosphere and dissolving in the sea means that it no longer acts as a greenhouse gas and does not contribute to further climate warming, thus mitigating atmospheric and terrestrial effects of rising CO_2. But this benefit of the 'marine carbon sink' comes at a cost. In seawater, CO_2 forms carbonic acid, increasing ocean acidity and making conditions less favourable for organisms depositing mineralized skeletons of calcium carbonate. The whole process is referred to as 'ocean acidification'.

Providing a palaeontological perspective, Charlie Veron noted that four of the five mass extinctions on earth dating back more than 400 million years were associated with rapidly rising or peaking levels of atmospheric CO_2. The losses in marine biodiversity following some of those extinctions were marked by 'reef gaps' (not only *coral* reef gaps, and not always associated with mass extinctions), each extending for several million years. When other potential causes are eliminated, 'the carbon cycle is the only player in the game big enough to inflict mass destruction on all of the Earth's terrestrial and marine life simultaneously.'[40]

Wolfgang Kiessling, however, argued that no single factor appeared to explain the waxing and waning of reefs, and that 'Ocean acidification as a single cause for the extinction-related reef crises is therefore unlikely.' While concluding that other stresses such as rapid climate change, excessive nutrification, oxygen shortage and hydrogen sulphide poisoning were also involved, he did allow that the 'possibility that ocean acidification played a role in some reef crises is certainly worth pursuing'.[41] It is this prospect that has driven contemporary research on the effect of ocean acidification on corals in an era of record rapid increases in seawater CO_2 and temperature, and their contributions to recent and projected future declines of coral reefs (illus. 217).

The concentration of CO_2 in the atmosphere had been fairly stable – between 172 and 300 parts per million (ppm) by volume – for the past 800,000 years, but since the onset of industrialization in the mid-eighteenth century it has increased rapidly to reach 400 ppm in 2015 and to exceed 405 ppm in 2017. Most of this increase has come in the past century, when the burning of fossil fuels added 30–40 times more CO_2 to the atmosphere than during the 150 years from 1750 to 1900.[42] Depending on several scenarios in which the burning of fossil fuels continues unabated ('business as usual') or is dialled back under the Paris Agreement, atmospheric CO_2 could approach 1,000 ppm by the year 2100.[43] The past increases correspond to a 30 per cent surge in ocean acidity. Under business as usual, acidity would triple by 2100, with potentially catastrophic effects for the most sensitive marine calcifiers, including calcareous plankton, molluscan shellfish and corals. This scenario, and global effects that already are manifest, helped to catalyse the Paris Agreement.[44]

The relationships between dissolved CO_2, acidity and calcification reside partly in the carbonate chemistry of seawater, carbonate (CO_3^{2-}) being the ion that combines with the calcium ion (Ca^{2+}) to precipitate as the skeletal mineral calcium carbonate ($CaCO_3$) – in chemical notation, $Ca^{2+} + CO_3^{2-} \rightarrow CaCO_3$. These chemical relationships are integrated as the 'saturation state' of calcium carbonate (aragonite in the case of scleractinian corals) in seawater, largely a function of the availability of carbonate because calcium is plentiful but carbonate diminishes and can become limiting as acidity increases. For reef-building scleractinians, the optimal value of the aragonite saturation state is above 4 (that is, four times the saturation concentration of carbonate in seawater); 2–3 is marginal; lower than that greatly impairs calcification; and below 1, $CaCO_3$ dissolves. Diverse corals have different physiological abilities to control the saturation state adjacent to their tissues where $CaCO_3$ is precipitated, by using cellular pumps to adjust ionic conditions and pH there, albeit at an energetic cost to the coral of pumping ions such as Ca^{2+} and H^+.[45] But ultimately, corals cannot contravene carbonate chemistry.

The issue is not only the levels of CO_2 and pH, but also how rapidly they are changing. The geological record indicates that the current rates are faster than at any time during the past 65 million years, and perhaps the last 300 million.[46] Indeed, in the end-Permian mass extinction about 250 million years ago, 'the most catastrophic loss of biodiversity in geological history', the disproportionate demise of heavily calcified marine organisms was driven by a rapid pulse of volcanic carbon dioxide release and ocean acidification (and the attendant stresses of warming and deoxygenation of seawater) comparable to our contemporary anthropogenic disturbance.[47] Under today's situation, given the inarguable increase in atmospheric carbon dioxide,

ocean acidification alone is likely to cause reef building to cease by the end of the 21st century... If coral bleaching due to ocean warming is also taken into account, then the rates of erosion on most reefs could outpace the overall reef building by corals and other organisms once CO_2 levels reach 560 ppm (by mid-century...)[48]

Will such scenarios be borne out? Natural submarine springs emitting lower-pH water, and volcanic CO_2 seeps that reduce the pH in the immediate area, provide real-world

217 The number of different types of reef sites (coral reefs indicated in red) during the past 500 million years, showing their waxing and waning. Vertical dashed lines indicate mass extinctions. Notice that not all reef gaps are associated with mass extinctions. The geological periods are abbreviated: Cm, Cambrian; O, Ordovician; S, Silurian; D, Devonian; C, Carboniferous; P, Permian; Tr, Triassic; J, Jurassic; K, Cretaceous; Pg, Palaeogene; N, Neogene.

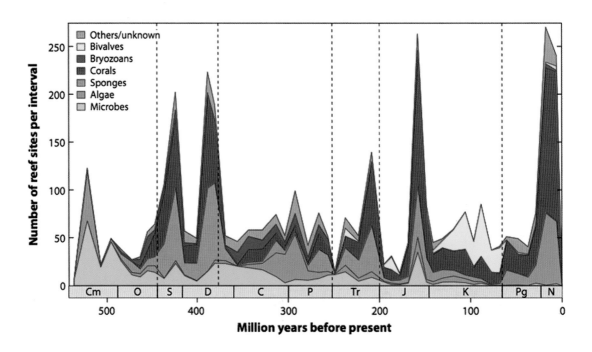

glimpses of the effects on corals and reefs of the ocean acidification forecast to occur by century's end. Four independent studies of widely separated sites in Papua New Guinea, the northern Caroline Islands, the Ryukyu Islands and Yucatán (Mexico) examined coral and community responses in shifting areas of CO_2, pH and saturation state, with seawater chemistry decreasingly conducive to calcification, compared with adjacent control areas (pH 8.1).[49] Broadly considered, the addition of CO_2 led to decreases in coral species diversity and loss of the reef-cementing calcareous red algae; increases in the proportion of massive or encrusting corals relative to branching or foliose corals (whose structural complexity offers greater habitat diversity for reef fishes and invertebrates); or takeovers by fleshy seaweeds and sea grasses (non-calcifiers forecast to do better in a high-CO_2 world) and alcyonacean soft corals. In some cases, calcification by reef corals decreased, producing skeletons that were less dense (imagine osteoporosis in corals) and more subject to erosion. Under extreme conditions in Papua New Guinea (pH 7.7, atmospheric equivalent of CO_2 = 980 ppm, aragonite saturation state = 2.0), reef development ceased (illus. 218, 219 and 220).

Yet in Palau, remarkably diverse hard coral communities have persisted for centuries inside the barrier reef and enclosed bays despite low levels of pH and saturation state resulting from poor flushing of CO_2 produced in respiration by the coral reef community.[50] Even in Palau, however, photographs of the reef sites show domination by colourful yet slow-calcifying, less complex massive and encrusting corals under lower pH and aragonite saturation (illus. 221).

On the flip side of studying naturally acidified sites, reversing several centuries of industrial ocean acidification by experimentally alkalinizing the seawater flowing across the reef flat at One Tree Island on the Great Barrier Reef saw the aragonite saturation state rise by 0.6 units – to nearly the pre-industrial level.[51] Net calcification by the alkalinized reef-flat community (corals, coralline algae and other calcifiers) likewise increased by 7 per cent compared to the unaltered control area. Thus, community calcification after the onset of industrialization may *already* have fallen by this same percentage as the saturation state decreased from 4.5 then to 3.8 now. Earlier studies of calcification by the reef community three decades apart (1975 and 2009) at Lizard Island on the northern GBR would suggest that nearly half of the post-industrial 7 per cent decline in net calcification occurred in recent decades when CO_2 levels were rising exponentially and saturation state declined from 4.3 to 3.9.[52] This timeline is consistent with the thirty-fold greater addition of CO_2 to the atmosphere from burning fossil fuels in the past century, compared to the 150 years from the onset of industrialization to 1900.

Collectively, these field studies and experiments provide a 'time machine' to visit coral reefs both in pre-industrial times and those foreseen for the end of the current century. Business as usual would see atmospheric CO_2 rise to above 900 ppm and saturation state decline to about 2, shifting the 2100 CE baseline to reefs that today are seen as degraded or dead. If the national pledges in the Paris Agreement are not renewed and strengthened in 2030, CO_2 will reach double the pre-industrial level (the oft-cited threshold for reef ruination) and beyond, perhaps 650 ppm; saturation state will fall to between 2.5 and 3,[53] and in surviving corals, calcification (already down by 7 per cent since the late 1700s, most of the decline coming in recent decades) will drop to about 50 per cent of what it was when James Cook navigated the Labyrinth.[54]

Reaffirming and strengthening the Paris pledges could hold atmospheric CO_2 below 500 ppm, with the aragonite saturation state declining less, to 3.5. Some corals may be able to

adapt, as those in Palau may have after living for generations under high CO_2, low pH and low carbonate saturation. The increasing dominance by hardy massive bouldering and encrusting corals would create a future with simpler reef ecosystems – structurally less complex and 'flattened', having fewer branching corals, with lower habitat variation and concomitantly reduced biodiversity. For example, demersal zooplankton shelter among corals during the day and emerge into the water column at night. The abundance of these zooplankton (a major food source for planktivorous reef fishes and invertebrates) is already lower at those naturally acidified reef sites in Papua New Guinea[55] where sheltering, complexly branched corals are being replaced by massive corals that offer fewer hiding places, thereby altering the food web of the reef community.

A worst-case scenario for what future reef sites might resemble if carbon emissions and ocean acidification were to continue unabated beyond this century emerged from an experiment in which corals were maintained either in ambient seawater (pH 8.1) or less alkaline seawater (pH 7.4, an extreme case). The skeleton dissolved at the reduced pH (because the concomitantly lower aragonite saturation state favoured dissolution, not precipitation) but the polyps survived; when returned to normal seawater, the polyps again secreted hard skeletons.[56] Something like this may have happened during 'reef gaps' in the geological record, and indeed it appears that the non-calcified close relatives of scleractinians – the corallimorpharians – evolved from reef corals that were unable to calcify under sub-saturating conditions.[57] But such 'corals' cannot build the massive structures that provide diverse habitats and make coral reefs embodiments of biodiversity, as well as affording 'ecosystem services' such as fisheries, tourist destinations and coastal protection from storms.

Despite early worries, the limited data indicate that calcification in two species of cold-water azooxanthellate scleractinian corals living deep in the Mediterranean is not depressed by

ocean acidification.[58] Perhaps during their long isolation in cold water, where the saturation state is lower than in warm water (because warm water holds more carbonate), deep-water corals have evolved more efficient mechanisms of calcification at low ambient saturation state.

Even so, their calcification rate is sufficiently slow that these cold-water corals will be unable to outgrow the pace of destruction of their reefs by bottom trawling for fish (illus. 222, 223 and 224). Despite a scientifically informed environmentalist campaign against such trawling,[59] in 2013 the EU voted to allow it to continue in the North Atlantic areas where the reefs occur. Eventually, increasing ocean acidity will promote the dissolution of the corals' exposed calcium carbonate framework, tipping the balance between construction and destruction.

Syndromes and Poxes and Plagues

In addition to thermal mass bleaching, deteriorating water quality and ocean acidification, microbial diseases increasingly threaten reef corals. First unwittingly illustrated in the nineteenth century,[60] and now rising in frequency and extent since the 1980s, outbreaks of diseases are becoming major sources of coral mortality, and the Caribbean in particular is a disease 'hotspot'. The word is overused but apt, both because more than 70 per cent of all reports of coral diseases are from this area,[61] and also because such outbreaks are associated with high-temperature anomalies, which are increasing on virtually all coral reefs and are exacerbated by agricultural runoff. Thermal stress lowers the corals' resistance to infection

218, 219 and 220 Volcanic CO_2 seeps in Papua New Guinea and their coral communities. (Left) Control area, low CO_2, pH ~8.1, with a diverse coral community; (Centre) high CO_2 area, pH 7.8–8.0, with mostly massive Porites *corals; (Right) extreme CO_2 area, pH 7.7, where algae and sponges have largely replaced corals. Notice the bubbles of CO_2 in the middle and right photographs.*

and sometimes increases the virulence of the pathogens. Caribbean 'white plague' of corals, caused by a coral-killing bacterium, increased after high-temperature bleaching events in Puerto Rico. Like human epidemics, coral diseases spread faster where coral populations are denser. Coral diseases seem more frequent on reefs close to large human populations, and the cause of the lethal 'white pox' that is decimating the Caribbean elkhorn coral *Acropora palmata* is a sewage-borne human faecal bacterium.[62]

More than thirty other 'syndromes' and diseases in corals have been described, but few of them meet long-established and modern diagnostic criteria to confirm a particular pathogen as the cause.[63] (illus. 225) In addition to white plague and white pox bacteria, infective agents include species of *Vibrio* that cause 'bacterial bleaching'

Above: 221 Diverse, colourful corals have survived in enclosed bays in Palau despite low values of pH and aragonite saturation state. Notice that most of the corals are massive or encrusting, not the branched forms that offer complex habitats for other animals and characterize more biologically diverse reefs. Opposite page: 222, 223 and 224 Mound of living colonies of Lophelia pertusa *on the Rockall Bank (top). Probable trawl damage, with fragmented* Lophelia *colonies scattered across the seabed (middle). Probable trawl damage showing an overturned boulder with trawl marks and a fragment of* Lophelia *in the background (bottom). The paired red dots are from a laser rangefinder and are 10 cm (3.94 in.) apart.*

of corals. The situation is complicated because in some cases a suite of many bacteria is associated with a disease. For example, in black band disease a zone of dying coral tissue spreads across the colony; accompanying this tissue necrosis is a consortium of fifty or more bacterial and fungal species that do not normally occur on corals. The mats of microorganisms rapidly consume the available oxygen and smother the corals and also release toxins.[64] Seawater holds even more viruses than bacteria, and these, too, can infect the coral host and its symbiotic algae. The usual suspects among environmental stressors increase the incidence of herpes-like viruses that infect the coral animal host.

Little is known about specific coral defences against these microbial and viral diseases. The layer of mucus secreted by the coral is home to huge numbers of bacteria that differ from those in the surrounding seawater and probably secrete anti-microbial chemicals that repel pathogenic boarders before they cause infection. In one case, corals susceptible to bleaching caused by an invading bacterium developed resistance to the pathogen by associating with other bacteria that conferred protection. Not only can corals recognize the 'not-self' tissues of competing corals and attack them, but they also have an immune system that protects against microbial invasion of wounds to colonies before they heal. This protection includes the aggregation of mobile cells that inactivate or encapsulate invaders such as *Aspergillus*, a fungus that infects gorgonian sea fans in the Caribbean. Spores of this terrestrial fungus enter seawater via runoff and even are deposited with dust from storms in the spreading Sahara, carried on an ill wind across the Atlantic.

Reef Recovery: Resilience, Resistance and Relocation

Coral reef ecologists increasingly speak of 'resilience', meaning the ability of reef communities (which comprise many species of corals and other reef inhabitants) to rebound from the impacts of major environmental changes, recovering their biological functions despite the imposed stresses. 'Resistance' refers more to the ability of populations or individuals of a particular species to tolerate environmental stresses. Both terms, and refuges afforded by marine protected areas (MPAs) where sheltered corals may serve as a source of larvae to restock and repair degraded reefs downstream, assume that some corals will be 'winners' in the ongoing struggle against a changing climate.

Still relevant in a new age of corals is the classic Darwinian concept of natural selection – whereby, based on their having advantageous characteristics or 'adaptations' that render them 'fit' in a changing environment, some individuals succeed and persist while others do not. The differences between such winners and losers are genetically based, a distinction that applies both to the endosymbiotic algae and their coral animal hosts.

The *process* of adaptation involves an evolutionary change in a population, often in response to an altered environment. This genetic change usually takes many generations, measured in centuries or millennia, and differences may accumulate in geographically separated populations that have low connectivity and limited exchange of genes.

The large difference in thermal resistance, seen as different bleaching thresholds, in corals from different geographic areas (for example, ~30°C in GBR corals versus >35°C in Arabian/Persian Gulf corals: see illus. 215) may be an example of local genotypic adaptation to their different thermal environments as the Gulf warmed over the past 6,000 years or so. It is not known, however, whether the thermal hardiness of the Gulf corals is attributable to the metabolism of resistant hosts or to their harbouring an especially tolerant species of symbiotic dinoflagellate, *Symbiodinium thermophilum*.[65] Bathymetric conditions

225 Five coral diseases for which the formal criteria to establish a particular pathogen as the cause of the disease have been fulfilled.

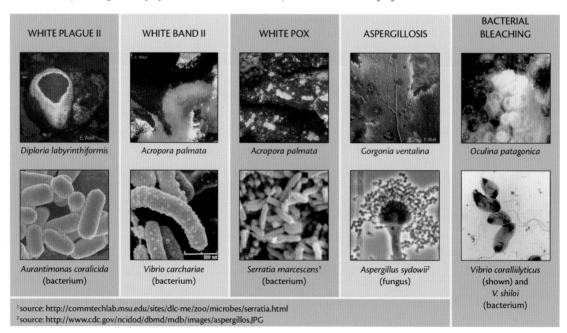

WHITE PLAGUE II	WHITE BAND II	WHITE POX	ASPERGILLOSIS	BACTERIAL BLEACHING
Diploria labyrinthiformis	*Acropora palmata*	*Acropora palmata*	*Gorgonia ventalina*	*Oculina patagonica*
Aurantimonas coralicida (bacterium)	*Vibrio carchariae* (bacterium)	*Serratia marcescens*[1] (bacterium)	*Aspergillus sydowii*[2] (fungus)	*Vibrio coralliilyticus* (shown) and *V. shiloi* (bacterium)

[1] source: http://commtechlab.msu.edu/sites/dlc-me/zoo/microbes/serratia.html
[2] source: http://www.cdc.gov/ncidod/dbmd/mdb/images/aspergillos.JPG

promoting the adaptation to high temperature and acidity by those corals in poorly flushed Palauan bays have developed during the past 150–500 years.[66]

Physiological 'acclimatization' involves short-term adjustments or compensations for environmental changes in individuals – of the order of days or weeks to months – that do not involve genetic changes. An example of acclimatization is the reversible seasonal change in the thermal bleaching threshold in a local population of corals (or their algae), being higher in summer than in winter. A coral's capacity for acclimatization may extend slightly beyond its normal annual range of temperature, for example when exposure to the long-term maximum temperature in an area induces innate (genetically determined) metabolic and cellular responses to heat stress that afford the coral the ability to withstand, albeit temporarily, abnormally higher temperatures. Ultimately, however, climate-related changes in local temperature will be so large that the physiological plasticity of acclimatization will be inadequate, because its extent or capacity is limited by the organism's (algal or animal) genetic endowment.

A case in point may be seen in the 27 years of sea surface temperature (SST) records for the Great Barrier Reef, which suggest 'physiological preparedness' (acclimatization or conditioning) stemming from sub-lethal stress and underlying historical spatial and temporal patterns of bleaching there.[67] Recall that most corals live within a degree or two of their bleaching or lethal threshold. In 75 per cent of the GBR records of thermal stress events, the thermal trajectory exceeded the monthly long-term (mean) maximum temperature for approximately two weeks and then transiently declined during a short recovery period, before rising again and exceeding the local bleaching threshold (benchmarked as the mean maximum +2°C). In such a case, bleaching would not occur because pre-exposure of the corals to a temperature above the local mean maximum but below the bleaching threshold induced their defences against a transiently higher temperature that would otherwise cause bleaching.

When (in 20 per cent of the cases of thermal stress) SST rose steadily (without a brief recovery period) and eventually exceeded the local bleaching threshold, corals were not protected and a single bleaching event occurred. A third trajectory (5 per cent of cases) exceeded the local bleaching threshold in two successive peaks, causing repetitive bleaching.

As SST continues to climb in the coming decades and the monthly maximum baseline shifts upward, a 0.5°C increase will see reefs that historically have experienced only the 'protective trajectory' begin to bleach repetitively. With fewer reefs experiencing the sub-bleaching acclimatization or conditioning that confers protection and has mitigated historical bleaching events, an overall decline in coral health, increase in repetitive bleaching mortality and a decrease in coral cover is forecast.[68] Such repetitive bleaching without adequate time for recovery is a probable death sentence for susceptible losers and an advent of tough times for quondam winners.

Mobile animals including fishes, insects, birds and mammals may migrate to more hospitable climes as global change shifts the range suitable for them, such as moving poleward as lower latitudes become too warm. Examples of range shifts have also been documented for individual species of corals, but the idea that entire coral reef ecosystems consisting of thousands of species of corals and associated organisms (illus. 226) can move to higher latitudes at rates sufficient to escape the pace of tropical warming is untenable. Some simple calculations suggest why.

Over a 1,500-km (930-mi.) north-to-south stretch of the GBR, average seawater temperature decreases southward by 2°C. If corals are locally adapted to these temperatures (and allowing for some degree of acclimatization as conditions change), to match even the optimistic rate of warming from climate change of 2°C over the next century, the entire ecosystem would have to travel southward by 1,500 km during the next hundred years, averaging 15 km (9 mi.) per year, to stay within its thermal comfort zone.[69] Some species having planktonic larvae that can cross such distances before settling down might be able to relocate, but the perhaps 10 per cent of zooxanthellate scleractinians having brooded larvae that settle close to their sessile parents would not. So, some species can relocate at the required rate, but many cannot, and it takes hundreds or thousands of species hundreds to thousands of years to build a coral reef.

Moving away from the equator would subject reef-building corals (which depend heavily on the products of photosynthesis provided by their zooxanthellae) to reduced daily amounts of sunlight in winter because of fewer hours of sunlight and a lower solar angle. Eventually this light limitation of growth would restrict corals to progressively shallower, brighter depths (by about 0.6 m/2 ft per degree of latitude), where they would be more subject to damage by waves, extremes of temperature and salinity, competition with seaweeds, increasing exposure at low tide,[70] and the adverse effects of solar UV-B radiation forecast to increase dramatically in the tropics after mid-century.[71] Moving northward away from the equator, as some Japanese tropical/subtropical corals *are* doing at the speed of 14 km (8 ½ mi.) per year, is not a long-term solution: this displacement will become limited by ocean acidification, which is progressing southward from higher latitudes, 'sandwiching' the corals between the stresses of decreasing carbonate availability and higher temperature.[72]

To be sure, reef ecosystems *have* shifted their geographic distributions over millennia as local conditions changed, but not at speeds sufficient to keep pace with today's projected high rate of climate change, and particularly in view of the other environmental factors – ocean acidification and rising sea levels – that will occur simultaneously. Human-assisted migration, such as wholesale transplantations of warm-adapted reef communities to distant areas in the process of warming, not only is difficult and expensive but also carries the ethical baggage associated with introduced or invasive species. Less risky would be the 'genetic rescue' of local populations of a species by adding genes from more robust populations, perhaps by enhancing reproductive connectivity through assisted colonization,[73] that is, transplanting resistant corals to interbreed with local populations in adjacent threatened areas. So, is there any chance that with human assistance corals can adapt quickly enough to keep pace with climate change? The rescue of one population by introducing genes from another, more naturally resistant, population is a type of rapid evolutionary adaptation that might occur in nature or through assisted colonization as part of reef restoration efforts. Growing concerns about the fate of corals (the canaries in our climatic coal mine) have accelerated research to identify less conventional possibilities.

Field studies and molecular analysis of *Acropora hyacinthus* in American Samoa revealed that colonies in the same reef system reciprocally relocated between a moderately variable habitat and a highly variable habitat (which regularly experienced temperatures above the bleaching threshold) evinced not only reversible acclimatization to high temperature but also a persistent increase in the expression of genes in the coral host associated with resistance to heat stress. This expression (that is, the appearance of an observable characteristic attributed to a particular gene or set of genes) was maintained even when the corals were returned to the less stressful habitat.[74]

The mechanism of such sustained gene expression associated with physiological

226 U.S. postage stamps collectively depicting a Pacific coral reef near Guam. Painting by John D. Dawson, 2003.

acclimatization may be found in the realm of epigenetics (currently a hot topic also in human development and disease), whereby molecules binding to the chromosome affect gene expression without changing the genetic sequences in the DNA – that is, without inducing mutations. The adaptive variation is already present in the genome, requiring only the environmental stimulus and epigenetic regulation to be expressed. It is not yet (as of 2017) known if the adaptive gene expressions in *A. hyacinthus* are heritable, or if this form of adaptation occurs in other species of corals, and thus whether the results are relevant to maintaining reef biodiversity as climate changes.[75] Attention is turning to the corals' changing microbiome as a mediator of the epigenetic effects,[76] including transmission of physiological plasticity to the next asexual or sexual generation.

Central to 'climate change physiology' is the importance of genetic variation for adaptive physiological responses to environmental stress. The recognition of thermally resistant clades of endosymbiotic algae is one example of how differential physiological success is being used to identify candidates for the 'coral arks' of reserves or MPAs to foster reef resilience.

In solitary animals, genetic variation is spread among individuals in the entire population, each experiencing and responding to its environment separately. As composite organisms, coral colonies consist of thousands of more closely linked, potentially genetically identical individual polyps that will show the same response to a stress. But what if a beneficial somatic mutation occurred in one polyp in the colony and was transferred to its daughter polyps during vegetative budding? Then the colony would become a mosaic of polyps that are not genetically identical members of the same clone but rather are bearers of multiple genotypes (in this case two) that may differentially affect their response to environmental stress, perhaps enhancing the survival of some members of the colony even if others perish. The incidence of somatic mutations in an average-size branching colony of *Acropora* is estimated to be perhaps 100 million,[77] affording a realistic likelihood that some of them will be beneficial and contribute disproportionately to colony growth. A study of several species of *Acropora*, *Pocillopora* and *Porites* detected an overall incidence of mosaicism in 31 per cent of the colonies tested.[78]

Such genetic mosaicism of the animal moiety of the colony may also extend to its simultaneously hosting different clades of symbiotic algae, which, as we have seen, afford different chances for adaptive responses to a changing environment. The colony includes winners and losers among its constituent polyps and their endosymbionts, but persists as its composition changes through differential survival. Visual evidence suggesting the operation of such a system can be seen in the patchy bleaching and partial mortality of stressed colonies.

The ancient view of coral as a mythical chimera has a modern rebirth relevant to intra-colony genetic variation: the occurrence of natural chimeras formed of the component parts of two different individuals. Gregariously settling planula larvae and internally brooded or even planktonic embryos of corals may fuse, with the resultant 'novel entity' that is an eventual polyp having cells and genes from both larvae. This greater repository of genetic variability in one individual, akin to the collective mosaic variability in a colony, opens the possibility of the polyp's changing the proportion of its somatic cellular constituents to present the appropriate response to environmental challenges. As such, Baruch Rinkevich sees coral chimerism 'as a novel tool to combat impacts of global change'.[79]

With the awareness of a climate changing faster than corals can naturally adapt,[80] and the growing recognition of the benefits afforded by healthy reefs to human societies,[81] coral reef conservation has been driven through insights offered both by selective breeding and molecular genetic technologies to considering human-assisted evolution – a new version of 'intelligent design' that is essentially coral eugenics – 'to augment the capacity of reef organisms to tolerate stress and to facilitate recovery after disturbances'.[82] Adaptive variation in the case of the endosymbionts might be boosted by inducing genetic mutations in cultures of millions of algae through applying a relevant stress (say, high temperature) in the laboratory and selecting algae bearing mutations that confer greater heat resistance. Alternatively, chemical mutagens or UV- or X-irradiation might greatly increase the rate of random mutations from which 'hopeful monsters' that bear profoundly mutated genotypes fortuitously beneficial in a changed climate could be selected and somehow placed in resistant hosts (themselves the hoped-for products of longer-term breeding programmes). Not everyone buys into the techno-fix, thinking it 'implausible that we're going to succeed in doing in a couple of years what evolution hasn't succeeded at over the past few hundred million years'.[83] But a leading proponent of human-assisted evolution in corals feels in an increasingly desperate situation 'that we're at this point where we need to throw caution to the wind and just try'.[84] The necessary public dialogue is just beginning.

227 Jason deCaires Taylor, Grace Reef, *2006, cement cast.*

Reef Restoration

The contemporary artist Jason deCaires Taylor has used the compelling image of human–coral transformation in his underwater installations in the Caribbean. There he submerged neutral-pH cement casts of human bodies and faces, to become overgrown by corals and other reef organisms in areas devastated by a hurricane that left only loose rubble and unstable sand, where neither the biblical wise man in the Book of Matthew nor coral recruits would build their home. The video and still images are at once arresting and disturbing, but the gradual encrustation and transformation appears ultimately hopeful of the regenerative power of coral reefs, so long as thermal and acidification stresses permit (illus. 227). The artist also collaborated with the Reef Ball Foundation, a non-profit organization that seeks to rehabilitate marine reefs, coral and otherwise, by deploying engineered modular 'reef balls' offering massive substrata on which larvae might settle or where coral fragments might be transplanted.

Eternal Reefs, Inc., part of the Reef Ball family of companies, goes further and offers a 'green burial' whereby families may cast the cremated remains of loved ones in environmentally friendly concrete reef balls, sometimes pre-seeded with living reef organisms, and place them in suitable areas, eventually to become overgrown by coral and thus part of the reef itself. As of old, corals are born of death, but here a New Age death that recalls Captain Nemo's musing about a peaceful tomb of coral (illus. 228).

These cultural examples are part of the larger realization that when conservation fails and reefs are degraded, or when natural disasters strike, assisted rehabilitation or 'restoration' may repair or re-establish reefs on local scales, akin to terrestrial reforestation. More controversial has been the use of Biorock™ technology, pioneered in the 1970s by the engineer Wolf Hilbertz and the reef ecologist Thomas J. Goreau. A large metal frame acted as a cathode when low-voltage direct current was passed through it, depositing calcium carbonate electrolytically. The self-repairing limestone structures grew abiotically, providing shelter to fishes, lobsters and other reef animals, and formed stable breakwaters that protected shorelines. Fragments of coral (especially those rescued from damaged reefs) could be transplanted onto the frame, where (and this was the controversy) they survived better and healed and grew faster than controls on un-electrified frames, albeit by unknown physiological mechanisms (illus. 229).[85]

The successes of Biorock reefs are all the more remarkable because they often start with rescued pieces of damaged coral colonies, which otherwise often show poor survival. Indeed, both low survival rates and the damage to donor reefs are arguments against using direct transplants to try to restore damaged reefs more widely. Instead, 'coral gardening' involves rearing small coral fragments in special nurseries, including floating ropes and trays suspended in mid-water, off the bottom in sheltered areas. Once the protected fragments have healed and grown larger away from competition and interference by established colonies and predators, they can be placed into reef areas needing remediation.[86] The technique of propagating corals by fragmentation (reminiscent of the sea nymphs' using the original branches transformed by Medusa as scions to populate the sea with red coral) and gardening has long been used in the aquarium trade, with the 'frags' being sold to hobbyists to stock their reef aquaria. Large public aquaria maintain and propagate stocks for reef-tank displays and scientific research (illus. 231). The fragments used to stock *in situ* nurseries also might be taken from local reefs, with an eye towards preserving particular sorts of reef communities.

For Rinkevich, 'the restored reefs of tomorrow will be different from the current or past reefs' because coral gardening and

228 An Eternal Reef memorial with flourishing marine life. Overleaf: 229 A Biorock™ structure used in a reef restoration project at Pemuteran Bay in Bali, Indonesia.

restoration increasingly look not backward to rebuild the past but rather forward, considering climate change and using corals whose genetic endowments will favour them under projected environmental changes.[87] The nurseries of a brave new world of corals may well render robust 'reservoirs of resilience' (illus. 232), but we must be mindful that although the corals of the Arabian/Persian Gulf already are adapted to high temperature, their diversity is lower than in contiguous costal areas of the western Indian Ocean and the Red Sea (see illus. 17).

Stephen Palumbi and his colleagues envisioned complementing their basic research on those hearty Samoan corals with a restoration project in which they would 'handpick the hardiest, fastest-growing and most heat-resistant corals for their smart [designer] reef' and compare it over time with a reef built using randomly chosen corals.[88] Together with the results of the reef restoration experiment, learning

whether the corals' toughness proves to be heritable (as shown for thermal tolerance in larvae of *Acropora millepora*[89]) may help foretell what reef ecosystems will look like in the oceanic 'hot and sour soup'[90] in the New Age of Corals – diverse arrays of winsome winners, or, even if not zombified, structurally homogeneous assemblages of hard-case holdouts?

The larger natural experiment is already in progress throughout the tropics. Beginning in the western Pacific in mid-2014 and Hawaii later that year, and reaching American Samoa by early 2015 during the hottest year on record,[91] the third global coral bleaching event was protracted into 2016 by a 'Godzilla El Niño' to become the longest ever during a still-hotter record year,[92] with devastating effects on corals throughout the Pacific.

Despite the recurrent bleaching on the Great Barrier Reef in 1998, 2002 and 2016,

Above: 230 Batfish cluster under a Biorock™ structure at the Pemuteran Bay restoration site. Opposite page: 231 Coral reefs and coral research are increasingly in the political arena. Professor Denis Allemand, Scientific Director of the Centre Scientifique de Monaco (left) explains the cultivation of corals for research to French President François Hollande (centre) and Prince Albert II of Monaco (near right) during a visit in 2013.

there is no indication of a protective effect of past bleaching (through acclimatization or adaptation) because the reefs and coral taxa that showed high bleaching in 1998 and 2002 also bleached severely in 2016.[93] Even those corals that were winners during moderate bleaching events joined the ranks of losers under more severe conditions. Just as worrisome is that better water quality and reduced fishing pressure in protected zones did not ameliorate the unprecedented bleaching in 2016. Although such protection might bolster resilience during less severe repetitive bleaching episodes, the authors offer the message that considering the increasing frequency of severe bleaching, 'securing a future for coral reefs' urgently demands reducing the rate of global warming. Less than a week before this assessment appeared in *Nature* in March 2017, the Great Barrier Reef Marine Park Authority reported that mass bleaching was occurring for an unheard-of second consecutive year.[94]

Almost simultaneously, the International Energy Agency offered on 17 March 2017 the hopeful news that global CO_2 emissions had remained flat at 32.1 gigatonnes per annum from 2014 to 2016 while the global economy grew more than 3 per cent during that period.[95] This decoupling of carbon emissions and economic growth suggests that slowing contributions to global climate change need not come at the expense of the world economy. However, the hiatus in emissions growth was not enough to put the world on course to keep the global temperature rise below 2°c, and further reductions are necessary. In the U.S., CO_2 emissions actually fell by 3 per cent as more efficient natural gas (much of it from controversial hydraulic fracturing, or 'fracking') and renewable energy sources such as wind replaced some coal burning.[96] Still, such optimistic news was not encouraging enough for President Trump, who barely a week later ordered the Environmental Protection Agency (EPA) to review the Clean Power Plan proposed by the Obama administration to help meet the U.S. commitment to the Paris Agreement. On 10 October 2017, EPA Administrator and climate-change denier Scott Pruitt repealed the Clean Power Plan, as expected.

Although CO_2 emissions had not risen in the previous three years, CO_2 once produced can persist for centuries, so atmospheric levels remained above 400 ppm, and NOAA's Coral Reef Watch reported that 2017 was the third consecutive year of mass bleaching on some reefs, a manifestation of global bleaching that was 'the longest, most widespread, and probably the most destructive ever recorded.'[97] Is this sequence a harbinger of the repetitive bleaching without respite that has been foretold to spell the demise of coral reefs?

Opposite page: 232 Prototype coral nursery suspended in open water at Eilat, Red Sea. The small, newly created fragments being handled by the diver gradually grow to form larger colonies (seen on the right) that can be transplanted onto hard substrata on the bottom. All fishes and invertebrates shown here have recruited from the plankton, developing the nursery into a floating artificial reef.

Coda: What Lies Ahead?

Philosophical and moral issues have very little place in the world today: this is the crisis point we have reached.
James Bowen, *The Coral Reef Era: From Discovery to Decline* (2015)

Coincident with the release in 2014 of the draft plan of *Reef 2050*, the Australian poet and activist Judith Wright's call to arms *The Coral Battleground* (1977) was republished in a new age of threat to the Great Barrier Reef. In her Introduction, Wright had quoted those lines from Kenneth Slessor's poem 'Five Visions of Captain Cook' (1931) that both remind us of coral's polysemic past and forewarn us of its fragility:

> Flowers turned to stone! Not all the botany
> Of Joseph Banks, hung pensive in a porthole,
> Could find the Latin for this loveliness,
> Could put the Barrier Reef in a glass box
> Tagged by the horrid Gorgon squint
> Of horticulture. Stone turned to flowers
> It seemed – you'd snap a crystal twig,
> One petal even of the water-garden,
> And have it dying like a cherry-bough.
> (illus. 233)

Wright also mused that 'if the Great Barrier Reef could think, it would fear us . . . We have its fate in our hands.' The writer Cory Doctorow applied this idea of a conscious, threatened reef in his online short story 'I, Row-Boat': 'The reef burned with shame that it had needed human intervention to survive the bleaching events, global temperature change.'

Thus, in the poetic and public perception as well as the political province in the New Age of Corals, it is less about 'corals' than 'coral reefs'.

233 Jörg Schmeisser, Under the Sea, Great Barrier Reef, *1977, colour etching.*

For Wright, the GBR was an environmental rallying point: with its rich biological diversity and historical and cultural meaning to both Aboriginal and immigrant Australians, it could inspire the informed public awareness and emotional connection necessary for bringing pressure to control exploitive environmental insults and to conserve the reef, and, not coincidentally, to save the planet. Yet tropical coral reefs are still mined for construction material and fill, smothered and befouled locally by coastal dredging, runoff and pollution, slimed by over-fertilized teeming algae, overfished, blasted or poisoned by illegal fishing, run upon by vessels, trod on by tourists, ransacked for souvenirs and to stock aquaria and ravaged by diseases. In her concluding pages Wright wrote of the ongoing struggle, relevant now as then:

> The Reef's fate is a microcosm of the fate of the planet. The battle to save it is itself a microcosm of the new battle within ourselves . . . and one in which the future of the human race may finally be decided.[1]

Barely two weeks after the bombings of Hiroshima and Nagasaki, Jean-Paul Sartre wrote for the first issue of *Les Temps modernes* of a future in which nuclear annihilation of the planet would be avoided only if humanity took responsibility for its (and the earth's) survival by *deciding* to live. The existential threat of a more insidious global environmental catastrophe was not yet apparent, but his words remain relevant in a time when humanity's commitment to life is just as much a determinant. For in the end, it is changes in the behaviour of individual humans (in a growing population of more than seven billion) and personal lifestyles (increasingly characterized by conspicuous overconsumption by its wealthiest members and growing consumerism in even the less well-off) that must prompt departure from business as usual (with its unbridled economic expansion and industrial glut) if the 'Anthrobscene' is to end other than with a whimper.[2]

Pope Francis recognized this necessity for change in his encyclical of 24 May 2015, 'On Care for Our Common Home'.[3] Feel-good, no-regrets activities such as reef restoration efforts at resorts may reflect positive shifts in some human attitudes and do raise awareness, bringing both alarm and hope. Hopeful, too, are the growing international governmental and private commitments to reef conservation, MPAs and the Paris Agreement (despite the Trump administration's withdrawal in 2017). But ultimately the root cause of global degradation or catastrophe is billions of humans; now, like so many polyps that built reefs worldwide, they must act collectively but in diminishing numbers to combat the fundamental problem of CO_2 emissions that threaten all as a consequence of the industrialized world's Faustian bargain.[4]

A review published in the front-line journal *Science* by a multidisciplinary group of marine experts in the run-up to COP21, the 2015 UN climate summit in Paris, aimed to educate policymakers and the public regarding the impacts specifically on the oceans and society of different scenarios of CO_2 emissions. Warm-water corals are but one example of groups already affected by the insalubrious synergy of ocean warming and acidification. The projection for most marine organisms is dire under business as usual.[5] Such forecasts led to the October 2015 call by the International Society for Reef Studies for a long-term reduction of atmospheric CO_2 to 350 ppm (from its current level of 400 ppm), which would require the near-term reduction of net CO_2 emissions to near zero and that 'most fossil fuels be left in the ground' if functional coral reef ecosystems are to be saved for posterity. Even if the stringent emissions controls needed to limit the global surface temperature increase to 2.7°C in the twenty-first century, or more optimistically to 2°C or even 1.5°C with more ambitious measures, are implemented,[6] the impact on corals will increase by 2100.[7]

Near the end of *A Reef in Time* (2008), Charlie Veron had written:

> This decade represents a window of choice, a period of crisis during which we must do whatever is necessary to prevent our century from becoming a transition to disaster. If we do not act, or do not act fast enough, marine and coastal ecosystems will be set on an irretrievable path of extermination that will eventually see a meltdown in coastal economies with devastating cost to natural environments and human societies.[8]

The papal encyclical, too, addressed ocean acidification and the pollution of coral reefs, and cast as a moral issue the wider need to address climate change, for the world is humanity's common home, and it is the world's poorest who will suffer the most. Pope Francis quoted the Catholic bishops of the Philippines, who as early as 1988 had wondered about their reefs, 'Who turned the wonderworld of the seas into underwater cemeteries bereft of color and life?' By that time the practice of fishing by explosives was well established on many a Philippine reef and a 'vital part of the place's economy' while being illegal nationally and vilified internationally. It is easy for well-off people to point and damn the destruction. James Hamilton-Paterson was of two minds and in *Playing with Water* (1987) offered a sympathetic view of a community of these fishers, recognizing 'the imperative of feeding . . . an entire family, of earning enough money to buy essentials', including medicine and education for the children. Nor was the destruction heedless,

> for theirs is no indiscriminate bombing. They discovered years ago that if their charges were too large or set to go off too deep they damaged the corals and that stocks of inshore fish depended entirely on healthy reefs.

But this mind was tempered by his recognition that

> the whole archipelago presents a gloomy prospect. No matter how skilled they may be the Filipinos are steadily destroying their corals . . . the detonations . . . [are] part of the increasing barrage sustained against the natural world under the often specious guise of feeding the massing mouths of humanity.[9]

Such ruin pales before the extent of bleached and moribund reefs in warming seas that are the consequence of a global economy from which the fishers derive little benefit.

Within recent generations we have already witnessed the demise of another iconic reef ecosystem built by and on prolific calcifiers. Driven nearly to extinction by human activities, these animals have been rescued here and there by human intervention in the form of mariculture and restoration not unlike what is being proposed for corals. For Rowan Jacobsen, in *A Geography of Oysters* (2008),

> The oyster reefs of the New World turned out to be the coral reefs of the temperate zone. Like the coral reefs, they were a skin of life atop a ledge of dead calcium carbonate. They harbored the same breathtaking diversity . . . And they were just as sensitive to destruction.

And they disappeared first, because people can't eat coral.[10]

Why, then, should people care about incomestible corals and their constructions? To be sure, the loss of the wild diversity of coral reefs (which contain 25 per cent of all known marine species despite reefs accounting for only one to two thousandths of the global sea surface area in aggregate, about the size of France) would be a calamity having unknown repercussions in the biosphere. But there have been many reef gaps before there were humans to document them. In an age of the commodification of nature and the monetization of everything, the 'ecosystem services' (commercial and traditional subsistence fishing, shoreline protection, tourism) provided by coral reefs, valued at $30 billion and benefiting 500 million people worldwide, are increasingly used to justify conserving them, and rightly so – human lives and indeed entire societies, not just coral reefs, are at stake.

There is more to it, however – something deeper and more fundamental in the human psyche. For Sir David Attenborough, quoted in *The Ecologist* in 2001, 'the overwhelming reason [for preserving nature] is Man's imaginative health,' increasingly recognized as an intrinsic ecosystem service.[11] For the biographer and social historian Richard Holmes, 'You could say that if our world is to be saved, we must understand it both scientifically and imaginatively.'[12]

More than any non-human inhabitant of rain forest, tundra, veldt, desert, sierra and even the abyss, diverse corals and the magnificent structures they build have figured in human culture – art, commerce, literature, medicine, music, mythology, philosophy, religion and science – for millennia, reaching back into prehistory. Failing to curtail CO_2 emissions to ameliorate climatic warming in the 'last, best chance' offered by the Paris Agreement would acquiesce in coral reefs becoming a faded baseline and the waning of coral's aesthetic, psychological and referential majesty. Such a failure, coming on the heels of the most prolonged and destructive global coral bleaching event ever registered – and despite encouraging early results in the selective breeding and assisted evolution of stress resistance in coral symbioses,[13] increasing successes in conservation and sustainability efforts,[14] the intentions of some states such as California and of u.s. corporations to keep their own pledges under the Paris Agreement despite the Trump administration's opting out, and a

Opposite page: 234 Edmund Dulac, 'And deeper than did ever plummet sound I'll drown my book', painted for Shakespeare's Comedy of The Tempest with Illustrations by Edmund Dulac *(1908), watercolour.*

heartening hiatus in the rise of global carbon emission rates from 2014 to 2017 – also would admit a wider defeat in the struggle to avoid global environmental catastrophe, possibly allowing our civilization's barely-open window of opportunity and choice to close (illus. 234).

Appendix: Maps Showing Locations Mentioned in the Text

Map 1 Mediterranean Sea and surrounding areas.

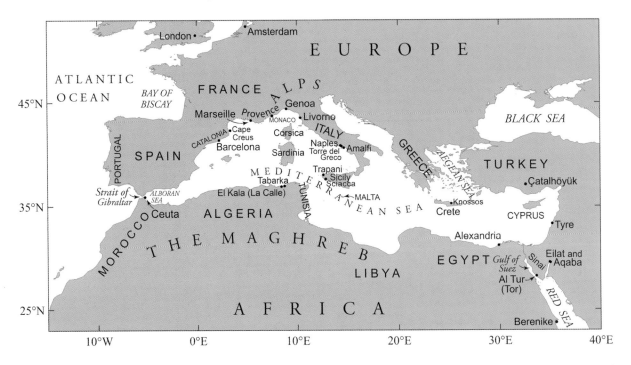

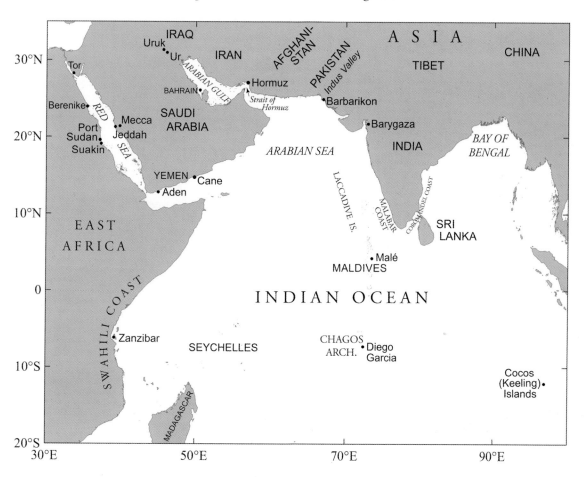

Map 2 Indian Ocean and surrounding areas.

Map 3 Western Pacific and Eastern Indian Oceans and surrounding areas.

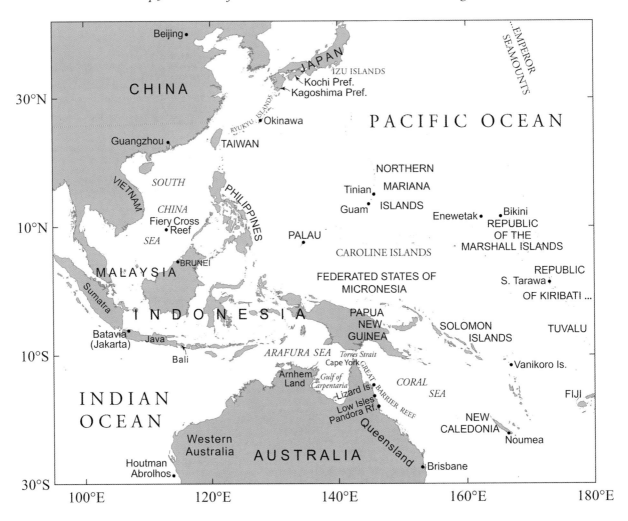

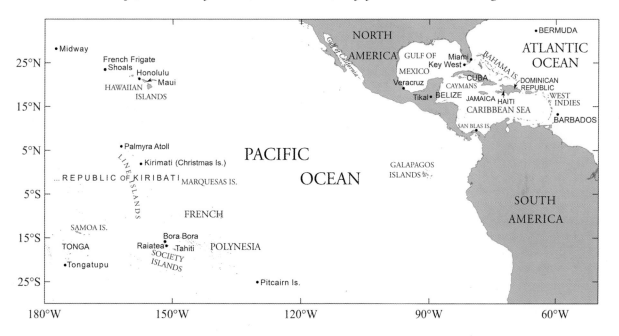

Map 4 Eastern Pacific Ocean, Caribbean Sea, Gulf of Mexico and surrounding areas.

References

Prelude: Where Corals Lie
1 Gustave Flaubert, *La Tentation de Saint Antoine* [1849–74], Project Gutenberg ebook #10982, www.gutenberg.org, 8 February 2004. All unattributed translations are by the author.
2 Barbara Maria Stafford, 'Picturing Ambiguity', in *Good Looking: Essays on the Virtue of Images* (Cambridge, MA, and London, 1996), p. 147.
3 R. W. Buddemeier, 'Making Light Work of Adaptation', *Nature*, CCCXXXVII (1997), pp. 229–30.
4 Marjorie L. Reaka-Kudla, 'Global Biodiversity of Coral Reefs: A Comparison with Rainforests', in *Biodiversity II: Understanding and Protecting Our Biological Resources*, ed. Marjorie L. Reaka-Kudla, Don E. Wilson and Edward O. Wilson (Washington, DC, 1997), pp. 83–108.

1 Defining Coral
1 François Poplin, 'Le Corail: entre animal, végétal, minéral et au cœur de la matière', in *Corallo di ieri, corallo di oggi*, ed. Marie-Laure Gamerre et al. (Bari, 2000), pp. 265–75.
2 M. Daly et al., 'The Phylum Cnidaria: A Review of Phylogenetic Patterns and Diversity 300 Years after Linnaeus', *Zootaxa*, 1668 (2007), pp. 127–82.
3 Heyo Van Ifen et al., 'Origin and Early Diversification of the Phylum Cnidaria Verrill: Major Developments in the Analysis of the Taxon's Proterozoic–Cambrian History', *Palaeontology*, LVII/4 (2014), pp. 677–90.
4 J. Malcolm Shick, *A Functional Biology of Sea Anemones* (London, 1991), p. 3.
5 Jian Han, Xingliang Zhang and Tsuyoshi Komiya, 'Integrated Evolution of Cnidarians and Oceanic Geochemistry before and during the Cambrian Explosion', in *The Cnidaria, Past, Present and Future: The World of Medusa and Her Sisters*, ed. Stefano Goffredo and Zvy Dubinsky (Cham, 2016), pp. 15–29.
6 E. Brendan Roark et al., 'Extreme Longevity in Proteinaceous Deep-sea Corals', *Proceedings of the National Academy of Sciences USA*, CVI/13 (2009), pp. 5204–8.
7 George D. Stanley Jr and Daphne G. Fautin, 'The Origins of Modern Corals', *Science*, CCXCI/5510 (2001), pp. 1913–14.
8 Mrs [Anna] Thynne, 'On the Increase of Madrepores', *Annals and Magazine of Natural History*, 3rd ser., 3 (1859), pp. 449–61 (p. 457).
9 Paulina Kaniewska et al., 'Signaling Cascades and the Importance of Moonlight in Coral Broadcast Mass Spawning', *eLife*, 4:e09991 (2015), DOI 10.7554/eLife.09991.
10 Robert Louis Stevenson's *Memoir* of Jenkin, quoted by Rosalind Williams in *The Triumph of Empire: Verne, Morris, and Stevenson at the End of the World* (Chicago, IL, 2013), p. 254.
11 J[oseph] Beete Jukes, *Narrative of the Surveying Voyage of HMS Fly, . . . in the Torres Strait, New Guinea, and Other Islands of the Eastern Archipelago, in the Years 1842–1846* (London, 1847), vol. I, p. 316.
12 Steve Jones, 'The Hydra's Head', in *Coral: A Pessimist in Paradise* (London, 2007), pp. 67–78.
13 Baruch Rinkevich and Yossi Loya, 'Senescence and Dying Signals in a Reef Building Coral', *Experientia*, XLII/3 (1986), pp. 320–22.
14 James Dwight Dana, *Corals and Coral Islands* (New York, 1872), p. 94.
15 Rinkevich and Loya, 'Senescence and Dying'.
16 Sandra Zielke and Andrea Bodnar, 'Telomeres and Telomerase Activity in Scleractinian Corals and *Symbiodinium* spp.', *Biological Bulletin*, CCXVIII/2 (2010), pp. 113–21; Hirotoshi Nakamichi et al., 'Somatic Tissues of the Coral *Galaxea fascicularis* Possess Telomerase Activity', *Galaxea, Journal of Coral Reef Studies*, XIV/1 (2012), pp. 53–9; Hiroki Tsuta et al., 'Telomere Shortening in the Colonial Coral *Acropora digitifera* during Development', *Zoological Science*, XXXI/3 (2014), pp. 129–34.
17 Dennis K. Hubbard, 'Reef Drilling', in *Encyclopedia of Modern Coral Reefs*, ed. David Hopley (Dordrecht, 2011), p. 863.
18 Anonymous, 'Coral Rings', in *Blackwood's Edinburgh Magazine*, LXXIV/455 (1853), pp. 360–71.
19 John W. Wells, 'Coral Growth and Geochronometry', *Nature*, 197 (1963), pp. 948–50; Colin T. Scrutton, 'Periodicity in Devonian Coral Growth', *Palaeontology*, 7 (1964), pp. 552–8, and plates 86 and 87.
20 Jules Rengade [Aristide Roger], *Voyage sous les flots*, 2nd edn (Paris, 1869), p. 161.
21 Walter M. Goldberg, *The Biology of Reefs and Reef Organisms* (Chicago, IL, and London, 2013), pp. 337–8.

22 J. Lang, 'Interspecific Aggression by Scleractinian Corals. 2. Why the Race is Not Only to the Swift', *Bulletin of Marine Science*, XXIII/2 (1973), pp. 260–79.
23 Georgyj M. Vinogradov, 'Growth Rate of the Colony of a Deep-water Gorgonarian *Chrysogorgia agassizi*: In Situ Observations', *Ophelia*, LIII/2 (2000), pp. 101–3.
24 S. Schmidt-Roach et al., 'Assessing Hidden Species Diversity in the Coral *Pocillopora damicornis* from Eastern Australia', *Coral Reefs*, XXXII/1 (2013), pp. 161–72.
25 M. Stat et al., 'Molecular Delineation of Species in the Coral Holobiont', *Advances in Marine Biology*, LXIII (2012), pp. 1–65.
26 George D. Stanley Jr, 'Photosymbiosis and the Evolution of Modern Coral Reefs', *Science*, CCCXII/5775 (2006), pp. 857–8.
27 Andrew C. Baker, 'Zooxanthellae', in *Encyclopedia of Modern Coral Reefs*, ed. David Hopley (Dordrecht, 2011), pp. 1189–92.
28 Forest Rohwer with Merry Youle, *Coral Reefs in the Microbial Seas* (Basalt, CO, 2010); Stat, et al., 'Molecular Delineation'; Tracy D. Ainsworth and Ruth D. Gates, 'Corals' Microbial Sentinels', *Science*, CCCLII/6293 (2016), pp. 1518–19; Margaret McFall-Ngai et al., 'Animals in a Bacterial World, a New Imperative for the Life Sciences', *Proceedings of the National Academy of Sciences of the USA*, 110 (2013), pp. 3229–36.
29 Kimberley A. Lema, Bette L. Willis and David G. Bourne, 'Corals Form Characteristic Associations with Symbiotic Nitrogen-fixing Bacteria', *Applied and Environmental Microbiology*, LXXVIII/9 (2012), pp. 3136–44.
30 Jeroen A.J.M. van de Water et al., 'Spirochaetes Dominate the Microbial Community Associated with the Red Coral *Corallium rubrum* on a Broad Geographic Scale', *Scientific Reports*, 6 (2016), DOI 10.1038/srep27277.

2 On the Nature of Corals

1 Eleni Voultsiadou and Dimitris Vafidis, 'Marine Invertebrate Diversity in Aristotle's Zoology', *Contributions to Zoology*, LXXVI/2 (2007), pp. 103–20.
2 G.E.R. Lloyd, 'Fuzzy Natures?', in *Aristotelian Explorations* (Cambridge, 1996), pp. 67–82.
3 Caroline Magdelaine, 'Le Corail dans la littérature médicale de l'Antiquité gréco-romaine au Moyen-Âge', in *Corallo di ieri, corallo di oggi*, ed. Marie-Laure Gamerre et al. (Bari, 2000), pp. 239–53.
4 Jean Théoridès, 'Consideration on the Medical Use of Marine Invertebrates', in *Oceanography: The Past*, ed. M. Sears and D. Merriman (New York, 1980), pp. 737–8.
5 Leo Wiener, *Contributions toward a History of Arabico-Gothic Culture*, vol. IV: *Physiologus Studies* (Philadelphia, PA, 1921), p. 178.
6 Gaius Plinius Secundus, *Historia naturalis*, Book XXXII, trans. John Bostock and H. T. Riley (London, 1855).
7 'Coral', in *Encyclopedia Iranica*, www.iranicaonline.org, 28 October 2011.
8 Roderich Ptak, 'Notes on the Word *Shanhu* and Chinese Coral Imports from Maritime Asia c. 1250–1600', *Archipel*, XXXIX/1 (1990), pp. 65–80 (p. 65).
9 Kenji Nakamori, 'Coral in Chinese Classics', in *A Biohistory of Precious Corals*, ed. Nozomu Iwasaki (Kanagawa, 2010), pp. 271–80.
10 Ibid., p. 272.
11 The quotations are from Boccone's 1671 *Recherches et observations curieuses sur la nature du corail blanc et rouge vray de Dioscoride*, given on page xi in H[enri] Milne Edwards, 'Introduction Historique', in *Histoire naturelle des coralliaires ou polypes proprement dits*, Tome Premier (Paris, 1857), pp. v–xxxiv.
12 H.M.E. de Jong, *Michael Maier's 'Atalanta fugiens': Sources of an Alchemical Book of Emblems* [1969] (Lake Worth, FL, 2014), p. 227.
13 Luigi Ferdinando Marsigli, Appendix II, 'Marsigli's Report to the Société Royale de Montpellier, 1706', in *Natural History of the Sea* [1725], trans. Anita McConnell, ed. Giorgio Dragoni (Bologna, 1999), pp. 19–21.
14 'The Voyage of Francois Pyrard de Laval, to the East Indies (an English-man being Pilot) and Especially His Observations of the Maldives, Where Being Ship-wracked Hee Lived Five Yeares. Translated out of French, and Abbreviated', in Samuel Purchas, *Hakluytus Posthumus, or Purchas His Pilgrimes* [1625] (Glasgow, 1905), vol. IX, Chapter 14, p. 509.
15 [Joseph Pitton le] Tournefort, 'Observations sur les plantes qui naissent dans le fond de la mer', Académie Royale des Sciences (1700), p. 34.
16 E. M. Beekman, trans. and ed., *The Ambonese Curiosity Cabinet: Georgius Everhardus Rumphius* (New Haven, CT, 1999), p. lxxx.
17 John Ellis, *The Natural History of Many Curious and Uncommon Zoophytes* (London, 1786), p. 146.
18 J[oseph] Beete Jukes, *Narrative of the Surveying Voyage of HMS 'Fly'* (London, 1847), vol. I, p. 119.
19 Edward Wotton, *De differentiis animalium libri decem* (Paris, 1552), p. 217.
20 [Georges-Louis Leclerc, Comte de] Buffon, Second discours. Histoire & théorie de la terre, article VIII, sur les coquilles & les autres productions de la mer, qu'on trouve dans l'interieur de la terre', in *Histoire naturelle, generale et particuliere, avec la description du Cabinet du Roy* (Paris, 1749), Book I, pp. 289–90.
21 Ibid., p. 290.
22 James Edward Smith, 'Letter from Linnaeus to Ellis, 16 September 1761', in *A Selection of the Correspondence of Linnaeus, and Other Naturalists, from the Original Manuscripts* (London, 1821), vol. I, pp. 151–2.

23 Samuel Taylor Coleridge, *Hints towards the Formation of a More Comprehensive Theory of Life*, ed. Seth B. Watson (London, 1848), p. 72.
24 Kathleen Coburn, *The Notebooks of Samuel Taylor Coleridge*, vol. I, Part 1: *1794–1804* (London, 1957), entry 841.
25 L'Abbé Dicquemare, 'Dissertation sur les limites des règnes de la nature', in *Observations sur la physique, sur l'histoire naturelle et sur les arts, avec des planches en taille-douce, dédiés à Mgr le comte d'Artois*, ed. l'abbé Rozier (Paris, 1776), vol. VIII, p. 376.
26 Gerardo Stecca, *Specious Morphology*, M.Sc. Thesis in Painting, The Savannah College of Art and Design (Savannah, GA, 2015), pp. 5–6, 31–3.
27 Daniel J. Boorstin, *The Discoverers* (New York, 1985), p. 444.
28 Karl A. Taube, 'Lidded Bowl with the Maize God in the Aquatic Underworld', in *Fiery Pool: The Maya and the Mythic Sea*, ed. Daniel Finamore and Stephen D. Houston, exh. cat. Peabody Essex Museum, Salem MA (Salem, MA, and New Haven, CT, 2010), p. 272.
29 James Edward Smith, 'Letter from Ellis to Linnaeus, 19 August 1768', in *A Selection of the Correspondence of Linnaeus, and Other Naturalists, from the Original Manuscripts* (London, 1821), vol. I, p. 230.
30 James Edward Smith, *Selection of the Correspondence of Linnaeus*, vol. I, p. 231.
31 James Bowen and Margarita Bowen, *The Great Barrier Reef: History, Science, Heritage* (Cambridge, 2002), p. 48.
32 Matthew Flinders, *A Voyage to Terra Australis; Undertaken for the Purpose of Completing the Discovery of that Vast Country, and Prosecuted in the Years 1801, 1802, and 1803, in His Majesty's Ship the 'Investigator'* (London, 1814), vol. II, p. 115.
33 Arthur Mangin, *The Mysteries of the Ocean*, trans. William Henry Davenport Adams (London, 1868), p. 182.
34 J.R.C. Quoy and J. P. Gaimard, 'Mémoire sur l'accroissement des polypes lithophytes considéré géologiquement', *Annales des sciences naturelles*, VI (1825), pp. 273–90.
35 D. R. Stoddart, 'Darwin, Lyell, and the Geological Significance of Coral Reefs', *British Journal for the History of Science*, IX/2 (1976), pp. 199–218, and David Dobbs, *Reef Madness: Charles Darwin, Alexander Agassiz, and the Meaning of Coral* (New York, 2005), give succinct accounts of this 'elevated-volcano' hypothesis.
36 Dobbs, *Reef Madness*, p. 152.
37 Jules Michelet, *The Sea* (New York, 1864), p. 125.
38 Ibid., p. 156.
39 T. H. Huxley, *A Manual of the Anatomy of Invertebrated Animals* (New York, 1878), p. 45.
40 Patrick Geddes, 'Further Researches on Animals Containing Chlorophyll', *Nature*, XXV (1882), pp. 303–5.
41 Frank Crisp, ed., *Journal of the Royal Microscopical Society; Containing its Transactions and Proceedings, and a Summary of Current Researches Relating to Zoology and Botany (principally Invertebrata and Cryptogamia), Microscopy, etc.* (1888), pp. 60–61.
42 W[illiam] Saville-Kent, *The Great Barrier Reef of Australia: Its Products and Potentialities* (London, 1893), pp. 157–8.
43 J. Stanley Gardiner, 'The Coral Reefs of Funafuti, Rotuma and Fiji Together with Some Notes on the Structure and Formation of Coral Reefs in General', *Proceedings of the Cambridge Philosophical Society*, IX (1898), pp. 417–503 (p. 484).
44 J. Stanley Gardiner, 'On the Rate of Growth of Some Corals from Fiji', *Proceedings of the Cambridge Philosophical Society*, XI (1901), pp. 214–19 (p. 215).
45 J. Stanley Gardiner, editorial footnote in C. A. MacMunn, 'On the Pigments of Certain Corals, with a Note on the Pigment of an Asteroid', in *The Fauna and Geography of the Maldive and Laccadive Archipelagoes*, ed. J. Stanley Gardiner (Cambridge, 1903), vol. I, part 2, p. 184 n. 1.
46 Bowen and Bowen, *The Great Barrier Reef*, chap. 15, and Barbara E. Brown, 'The Legacy of Professor J. Stanley Gardiner FRS to Reef Science', *Notes and Records of the Royal Society*, LXI/2 (2007), pp. 207–17.
47 C[harles] M[aurice] Yonge, *A Year on the Great Barrier Reef* (London, 1930), p. 111.
48 Gardiner, *The Fauna and Geography of the Maldive and Laccadive Archipelagoes*, vol. I, p. 422.
49 L. Muscatine and E. Cernichiari, 'Assimilation of Photosynthetic Products of Zooxanthellae by a Reef Coral', *Biological Bulletin*, CXXXVII/3 (1969), pp. 506–23.
50 Thomas F. Goreau and Nora I. Goreau, 'The Physiology of Skeleton Formation in Corals. II. Calcium Deposition by Hermatypic Corals under Various Conditions in the Reef', *Biological Bulletin*, CXVII/2 (October 1959), pp. 239–50, and Vicki Buchsbaum Pearse and Leonard Muscatine, 'Role of Symbiotic Algae (Zooxanthellae) in Coral Calcification', *Biological Bulletin*, CXLI/2 (October 1971), pp. 350–63.
51 Denis Allemand et al., 'Coral Calcification, Cells to Reefs', in *Coral Reefs: An Ecosystem in Transition*, ed. Zvy Dubinsky and Noga Stambler (Dordrecht, Heidelberg, London and New York, 2011), pp. 119–50.
52 J. Malcolm Shick, 'Why Don't Corals Get Sunburned?', www.institut-ocean.org/images/articles/documents/1434363042.pdf, June 2015.
53 A. Starcevic et al., 'Gene Expression in the Scleractinian *Acropora microphthalma* Exposed to High Solar Irradiance Reveals Elements of Photoprotection and Coral Bleaching', *PLOS ONE*, V/11 (2010), e13975.
54 J. M. Shick and W. C. Dunlap, 'Mycosporine-like Amino Acids and Related Gadusols: Biosynthesis, Accumulation, and UV-protective Functions in Aquatic Organisms', *Annual Review of Physiology*, LXIV (2002), pp. 233–62.

55 J. M. Shick and J. A. Dykens, 'Oxygen Detoxification in Alga–Invertebrate Symbioses from the Great Barrier Reef', *Oecologia*, LXVI/1 (1985), pp. 33–41; M. P. Lesser, 'Oxidative Stress in Marine Environments: Biochemistry and Physiological Ecology', *Annual Review of Physiology*, LXVIII (2006), pp. 253–78; Walter C. Dunlap, J. Malcolm Shick and Yorihito Yamamoto, 'Sunscreens, Oxidative Stress and Antioxidant Functions in Marine Organisms of the Great Barrier Reef', *Redox Report*, IV/6 (1999), pp. 301–6.
56 Baron Eugène de Ransonnet-Villez, *Sketches of the Inhabitants, Animal Life and Vegetation in the Lowlands and High Mountains of Ceylon, as Well as of the Submarine Scenery near the Coast, Taken in a Diving Bell* (London, 1867), p. 21.
57 Vincent Pieribone and David F. Gruber, *Aglow in the Dark: The Revolutionary Science of Biofluorescence* (Cambridge, MA, 2005), p. 87.
58 Charles H. Mazel et al., 'Green-Fluorescent Proteins in Caribbean Corals', *Limnology and Oceanography*, XLVIII/1:2 (2003), pp. 402–11.
59 Ibid., pp. 409–10.
60 J. M. Shick et al., 'Ultraviolet-B Radiation Stimulates Shikimate Pathway-dependent Accumulation of Mycosporine-like Amino Acids in the Coral *Stylophora pistillata* Despite Decreases in its Population of Symbiotic Dinoflagellates', *Limnology and Oceanography*, XLIV/7 (1999) pp. 1667–82.
61 A. Salih et al., 'Fluorescent Pigments in Corals are Photoprotective', *Nature*, CDVIII (2000), pp. 850–53.
62 Fadi Bou-Abdallah, N. Dennis Chasteen and Michael P. Lesser, 'Quenching of Superoxide Radicals by Green Fluorescent Protein', *Biochimica et Biophysica Acta*, MDCCLX/11 (2006), pp. 1690–95.
63 John Barrow, *A Voyage to Cochinchina in the Years 1792 and 1793* (London, 1806), p. 168.
64 E. P. Odum, 'How to Prosper in a World of Limited Resources: Lessons from Coral Reefs and Forests in Poor Soils', in *Ecological Vignettes: Ecological Approaches to Dealing with Human Predicaments* (Amsterdam, 1998), p. 95.

3 The Mythos, Menace and Melancholy of Corals
1 Ovid, *Metamorphoses*, trans. Charles Martin (New York, 2004), Book 4, ll. 1019–26.
2 Françoise Frontisi-Ducroux, 'Andromède et la naissance du corail', in *Mythes grecs au figuré: de l'Antiquité au Baroque*, ed. Stella Georgoudi and Jean-Pierre Vernant (Paris, 1996), pp. 135–65.
3 Lynn Thorndike, 'The Spurious Mystic Writings of Hermes, Orpheus, and Zoroaster', in *A History of Magic and Experimental Science* (New York, 1923), vol. I, p. 293.
4 The fourteenth-century Italian translation of the *Metamorphoses* was by Giovanni Bonsignore, not printed until 1497, and translated into English by Michael Cole in 'Cellini's Blood', *Art Bulletin*, LXXXI (1999), pp. 215–35 (p. 228).
5 G. Evelyn Hutchinson, 'The Enchanted Voyage: A Study of the Effects of the Ocean on Some Aspects of Human Culture', *Journal of Marine Research*, XIV (1955), pp. 276–83.
6 Akemi Iwasaki, 'The Language of Coral – the Vocabulary and Process of its Transformation from Marine Animal into Jewellery and Craftwork', in *A Biohistory of Precious Corals: Scientific, Cultural and Historical Perspectives*, ed. Nozomu Iwasaki (Kanagawa, 2010), p. 127.
7 Celeste Olalquiaga, *The Artificial Kingdom: A Treasury of the Kitsch Experience* (New York, 1998), caption to 'Treasures of the Sea', or 'Allegory of the Discovery of America', following p. 244.
8 'Gemology: The Mystery Hidden in Stones', http://solutionastrology.com/gemologydetails.asp?gemologyid=211, accessed 20 January 2014.
9 Massimo Vidale et al., 'Symbols at War: The Impact of *Corallium rubrum* in the Indo-Pakistani Subcontinent', in *Ethnobiology of Corals and Coral Reefs*, ed. Nemer E. Narchi and Lisa L. Price (Cham, Heidelberg, New York, Dordrecht and London, 2015), pp. 59–72 (p. 65).
10 'Hindu Astrology', http://en.wikipedia.org, accessed 20 January 2014.
11 Nitin Kumar, 'Color Symbolism in Buddhist Art', www.wou.edu, accessed 20 January 2014.
12 'Ryūjin', http://en.wikipedia.org, accessed 20 January 2014.
13 Nahoko Kahara, 'Momotaro and Precious Coral', in *A Biohistory of Precious Corals: Scientific, Cultural and Historical Perspectives*, ed. Nozomu Iwasaki (Kanagawa, 2010), pp. 251–70.
14 E.C.L. During Caspers, 'In the Footsteps of Gilgamesh: In Search of the "Prickly Rose"', *Persica*, XII (1987), pp. 57–95.
15 Ian S. McIntosh, 'Aboriginal Management of the Sea', in *Aboriginal Reconciliation and the Dreaming: Warramiri Yolngu and the Quest for Equality* (Boston, MA, 2000), p. 102.
16 R. Aldington and D. Ames, trans., 'Oceania Mythology. The Great Myths of Oceania', in *New Larousse Encyclopedia of Mythology* [1959] (New York, 1968), available at www.scribd.com.
17 Martha Beckwith, 'The Kane Worship', in *Hawaiian Mythology* (1940), pp. 42–59, at www.sacred-texts.com.
18 Toni Makani Gregg et al., 'Puka Mai He Ko'a: The Significance of Corals in Hawaiian Culture', in *Ethnobiology of Corals and Coral Reefs*, ed. Nemer E. Narchi and Lisa L. Price (Cham, Heidelberg, New York, Dordrecht and London, 2015), pp. 103–15.
19 'Opuhala', https://glitternight.com/2011/03/02/eleven-more-deities-from-hawaiian-mythology-2/, accessed 7 November 2017.

20 Jonathan Maberry and David F. Kramer, *They Bite* (New York, 2009), p. 231.
21 Martha W. Beckwith, trans. and commentary, *The Kumulipo: A Hawaiian Creation Chant* [1951], at www.sacred-texts.com.
22 Queen Lilioukalani, trans., *The Kumulipo* [1897], at www.sacred-texts.com.
23 Kenneth P. Oakley, 'Fossils Collected by the Earlier Palaeolithic Men', in *Mélanges de préhistoire, d'archéocivilisation et d'ethnologie offerts à André Varagnac* (Paris, 1971), pp. 581–4.
24 Kenneth P. Oakley, 'Emergence of Higher Thought 3.0–0.2 Ma BP', *Philosophical Transactions of the Royal Society of London, B*, CCXCII (1981), pp. 205–11.
25 Randall White, 'Technological and Social Dimensions of "Aurignacian-age" Body Ornaments across Europe', in *Before Lascaux: The Complex Record of the Early Upper Paleolithic*, ed. Heidi Knecht, Anne Pike-Tay and Randall White (Boca Raton, FL, 1993), pp. 286–7.
26 Kenneth P. Oakley, 'Fossil Coral Artifact from Niah Cave', *Asian Perspectives*, XX/1 (1977), pp. 69–74.
27 Alexander von Schouppé, 'Episodes of Coral Research up to the 18th Century', *Courier Forschungsinstitut Senckenberg*, CLXIV (1993), pp. 1–16.
28 S. A. Callisen, 'The Evil Eye in Italian Art', *Art Bulletin*, XIX (1937), pp. 450–62.
29 Samuel Purchas, 'The Voyage of Francois Pyrard de Laval, to the East Indies (an English-man Being Pilot) and Especially His Observations of the Maldives, Where Being Ship-wracked Hee Lived Five Yeares. Translated out of French, and Abbreviated', in *Haklutyus Posthumus, or Purchas His Pilgrimes* [1625] (Glasgow, 1905) vol. IX, pp. 508–9.
30 Patrick O'Brian, *Joseph Banks* (Boston, MA, 1993), p. 131.
31 Jean-René-Constant Quoy, 'On the Loss of the Lapérouse Expedition', in *An Account in Two Volumes of Two Voyages to the South Seas by Captain (later Rear Admiral) Jules S-C Dumont d'Urville*, vol. I: *Astrolabe, 1826–29*, trans. and ed. Helen Rosenman (Melbourne, 1987), p. 241.
32 Jules Verne, *Twenty Thousand Leagues under the Sea, the Definitive, Unabridged Edition Based on the Original French Texts*, trans. and annotated by Walter James Miller and Frederick Paul Walter (Annapolis, MD, 2003), pp. 141–2.
33 J.E.N. Veron, 'The Big Picture', *A Reef in Time: The Great Barrier Reef from Beginning to End* (Cambridge, MA, 2008), pp. 2–3.
34 James D. Dana, *Corals and Coral Islands* (New York, 1872), p. 19.
35 Captain Jules S.-C. Dumont d'Urville, 'Astrolabe at Vanikoro', in *An Account in Two Volumes of Two Voyages to the South Seas*, vol. I: *Astrolabe, 1826–1829*, trans. and ed. Helen Rosenman (Melbourne, 1987), pp. 210–40.
36 J[oseph] Beete Jukes, *Narrative of the Surveying Voyage of HMS Fly, . . . in the Torres Strait, New Guinea, and Other Islands of the Eastern Archipelago, in the Years 1842–1846* (London, 1847), vol. I, pp. 121–4.
37 Adam Gopnik, 'Darwin's Eye', in *Angels and Ages: A Short Book about Darwin, Lincoln, and Modern Life* (New York, 2009), p. 77.
38 T. A. Stephenson, 'Coral Reefs', *Endeavour*, V/19 (1946), pp. 96–106 (p. 105).
39 Richard Holmes, 'Shelley Undrowned', in *This Long Pursuit: Reflections of a Romantic Biographer* (New York, 2017), p. 247.
40 Jules Rengade [Aristide Roger], *Voyage sous les flots*, 2nd edn (Paris, 1869), p. 162.
41 Verne, *Twenty Thousand Leagues under the Sea*, p. 179.
42 Cory Doctorow, 'I, Row-Boat' (2006), at http://flurb.net/1/doctorow.htm.
43 Georges Cuvier, 'Litophytes', in *Discours sur les révolutions de la surface du globe, et sur les changements qu'elles ont produits dans le règne animal*, 3rd edn (1825), available at www.victorianweb.org, accessed 21 October 2014.
44 Hans Christian Andersen, 'The Little Mermaid', http://hca.gilead.org.il/li_merma.html, accessed 7 November 2017.
45 Shannon Kelley, 'The King's Coral Body: A Natural History of Coral and the Post-tragic Ecology of *The Tempest*', *Journal for Early Modern Cultural Studies*, XIV/1 (2014), pp. 115–42.
46 Jonathan Bate, *Shakespeare and Ovid* (Oxford, 2001).
47 Callum Roberts, *The Ocean of Life* (New York, 2012), chap. 3, 'Life on the Move', p. 86.
48 Jean-Georges Harmelin, Station Marine d'Endoume, Marseille, personal communication, email, 8 February 2012.
49 J.G.P. Delaney, *Glyn Philpot: His Life and Art* (Brookfield, VT, 1999), p. 105.
50 C. B. Klunzinger, *Upper Egypt: Its People and Products* (New York, 1878), p. 369.
51 J. Malcolm Shick, 'Otherworldly', in *Underwater*, exh. cat., Towner Art Gallery, Eastbourne, East Sussex (2010), pp. 33–9 (p. 38).
52 Quoted in Iain McCalman, *Reef: A Passionate History* (New York, 2013), p. 199.
53 Lester D. Stephens and Dale R. Calder, 'A Zeal for Zoology', in *Seafaring Scientist: Alfred Goldsborough Mayor, Pioneer in Marine Biology* (Columbia, SC, 2006), p. 16.

4 Conjuring Corals

1 Matthew Flinders, *A Voyage to Terra Australis; Undertaken for the Purpose of Completing the Discovery of that Vast Country, and Prosecuted in the Years 1801, 1802, and 1803, in His Majesty's Ship the 'Investigator'* (London, 1814), vol. II, pp. 87–8.
2 John Barrow, *A Voyage to Cochinchina in the Years 1792 and 1793* (London, 1806), p. 166.

3 Erasmus Darwin, 'The Economy of Vegetation', in *The Botanic Garden, a Poem, in Two Parts; Containing the Economy of Vegetation and the Loves of the Plants, with Philosophical Notes* [1791] (London, 1825), p. 44, Canto III, l. 90.
4 J. Malcolm Shick, 'Toward an Aesthetic Marine Biology', *Art Journal*, LXVII/4 (2008), pp. 62–86 (p. 72).
5 Philip Henry Gosse, *Actinologia Britannica: A History of the British Sea-anemones and Corals* (London, 1860), pp. 15–16.
6 Ursula Harter, 'Les Jardins Océaniques', in *Odilon Redon, Le Ciel, la terre, la mer*, ed. Laurence Madeleine, exh. cat., Musée Léon-Dierx, Saint-Denis, Réunion (Paris, 2007), p. 141.
7 J[oseph] Beete Jukes, *Narrative of the Surveying Voyage of HMS Fly, . . . in the Torrest Straits, New Guinea, and Other Islands of the Eastern Archipelago, in the Years 1842–1846* (London, 1847), vol. I, p. 117.
8 Ibid., pp. 117–18.
9 [Christian Gottfried] Ehrenberg, 'Über die Natur und Bildung der Corallenbänke des rothen Meeres', *Abhandlungen der Königlichen Akademie der Wissenschaften in Berlin* (1832), pp. 381–438 (p. 383).
10 André Breton, *Mad Love*, trans. Mary Ann Caws (Lincoln, NE, 1987), p. 11 (*L'Amour fou*, Paris, 1937, p. 14).
11 Ann Elias, 'Sea of Dreams: André Breton and the Great Barrier Reef', *Papers of Surrealism*, 10 (Summer 2013), pp. 1–15, at www.surrealismcentre.ac.uk.
12 Breton, *Mad Love*, pp. 11–12 (*L'Amour fou*, pp. 14–15).
13 Kenji Nakamori, 'Coral in Chinese Classics', in *A Biohistory of Precious Corals*, ed. Nozomu Iwasaki (Kanagawa, 2010), p. 272.
14 Charles Darwin, *Charles Darwin's Beagle Diary*, ed. Richard Darwin Keynes (Cambridge, 1988), journal p. 714; see also Richard Milner, 'Seeing Corals with Darwin's "Eye of Reason": Discovering an Image of a Tropical Atoll in the English Countryside', in *Ethnobiology of Corals and Coral Reefs*, ed. Nemer E. Narchi and Lisa L. Price (Cham, Heidelberg, New York, and London, 2015), chap. 2, pp. 15–25.
15 [Louis] Aragon, *Henri Matisse, roman* [1971] (Paris, 1998), vol. I, assembled from fragments on pp. 21–3.
16 R. B. Williams and P. G. Moore, 'An Annotated Catalogue of the Marine Biological Paintings of Thomas Alan Stephenson (1898–1961)', *Archives of Natural History*, XXXVIII (2011), pp. 242–66.
17 P[hilip] H[enry] Gosse, 'Multum e Parvo', in *The Romance of Natural History* [1862] (New York, 1902), p. 94.
18 Baron Eugène de Ransonnet, *Sketches of the Inhabitants, Animal Life and Vegetation in the Lowlands and High Mountains of Ceylon, as Well as of the Submarine Scenery near the Coast, Taken in a Diving Bell* (Vienna and London, 1867), pp. 21–2.
19 Stefanie Jovanovic-Kruspel, Valérie Pisani and Andreas Hantschk, 'Under Water – between Science and Art – the Rediscovery of the First Authentic Underwater Sketches by Eugen von Ransonnet-Villez (1838–1926)', *Annalen des Naturhistorischen Museums in Wien*, series A, 119 (2017), pp. 131–53.
20 Pritchard quoted in J. Malcolm Shick, 'Otherworldly', in *Underwater*, exh. cat., Towner Art Gallery, Eastbourne, East Sussex (2010), pp. 33–9, and Margaret Cohen, 'Underwater Optics as Symbolic Form', *French Politics, Culture and Society*, XXXII/3 (2014), pp. 1–23, DOI 10.3167/fpcs.2014.320301.
21 Zarh H. Pritchard, *Appreciations of the Work of Zarh H. Pritchard*, typed transcription of written comments by visitors to his exhibitions. Unpaginated and undated; received by the Musée Océanographique de Monaco in 1921.
22 Rosamond Wolff Purcell and Stephen Jay Gould, 'Dutch Treat: Peter the Great and Frederik Ruysch', in *Finders, Keepers: Eight Collectors* (New York and London, 1992), pp. 13–32.
23 Marie-Claude Beaud and Robert Calcagno, 'Curious!', in *Oceanomania: Souvenirs of Mysterious Seas from the Expedition to the Aquarium*, exh. cat., Nouveau Musée National de Monaco and MACK, London (2011), p. 19.
24 [Joseph Pitton de] Tournefort, 'Observations sur les plantes qui naissent dans le fond de la mer', *Mémoires de l'Académie Royale des Sciences* (13 February 1700), p. 36.
25 Gordon Williams, 'Coral Penis', in *A Dictionary of Sexual Language and Imagery in Shakespearean and Stuart Literature*, vol. I: *A–F* (London, 1994), pp. 306–7.
26 Ibid., p. 307.
27 Nakamori, 'Coral in Chinese Classics', p. 271.
28 Jules Michelet, 'Blood-Flower', in *The Sea* (La Mer) (New York, 1864 [1861]), pp. 147–8.
29 Émile Zola, *The Kill* [1871–2], trans. Arthur Goldhammer (New York, 2005), p. 241.
30 Lori Baker, *The Glass Ocean* (New York, 2013), p. 64.
31 Côme Fabre, 'Le Romantisme noir à l'heure symboliste: la perversité de Dame Nature', in *L'Ange du bizarre: le romanticisme noir de Goya à Max Ernst*, ed. Côme Fabre and Felix Krämer, exh. cat., Städel Museum, Frankfurt, and Musée d'Orsay, Paris (2013), pp. 147–53.
32 Aragon, *Henri Matisse, roman*, vol. I, pp. 23–4.
33 Horace Keats, 'The Coral Reef', in *Drake's Call: A Collection of Songs of the Sea for Low Voices/Music by Horace Keats*, ed. Wendy Dixon, David Miller and Brennan Keats (Culburra Beach, Australia, 2001), pp. 3–14.
34 Toru Takemitsu, *Coral Island (An Atoll)* [1962], quotes from the back cover of the RCA Victor album VICS-1334 (1968).
35 See https://stuart-mitchell.com.

5 Coral as Commodity
1 Masayuke Nishie, 'Precious Coral from a Cultural Perspective', in *A Biohistory of Precious Corals: Scientific,*

Cultural and Historical Perspectives, ed. Nozomu Iwasaki (Kanagawa, 2010), p. 102.
2 Maria A. Borrello et al., 'Les Parures néolithiques de corail (*Corallium rubrum* L.) en Europa occidentale', *Rivista di Scienze Preistoriche*, LXII (2012), pp. 67–82 (pp. 69–70).
3 Maria Angelica Borrello, 'Vous avez dit "corail"?', *Annuaire de la Société Suisse de Préhistoire et d'Archéologie*, LXXXIV (2001), pp. 191–6.
4 James Mellaart, 'Excavations at Çatal Hüyük, 1962: Second Preliminary Report', *Anatolian Studies*, XIII (1963), pp. 43–103.
5 Borrello et al., 'Les Parures néolithiques de corail'.
6 Sara Champion, 'Coral in Europe: Commerce and Celtic Ornament', in *Celtic Art in Ancient Europe: Five Protohistoric Centuries* (London and New York, 1976), pp. 29–37.
7 Malcolm H. Wiener, 'The Nature and Control of Minoan Foreign Trade', *Studies in Mediterranean Archaeology*, vol. XC, 'Bronze Age Trade in the Mediterranean', ed. N. H. Gale (1991), pp. 325–50.
8 Michel Glémarec, *Mathurin Méheut, décorateur marin* (Brest, 2013), p. 79.
9 Dimitri Meeks, 'Le Corail dans l'Égypte ancienne', in *Corallo di ieri, corallo di oggi*, ed. Jean-Paul Morel, Celia Rondi-Costanzo and Daniela Ugolini (Bari, 2000), pp. 99–117.
10 Franck Perrin, 'L'Origine de la mode du corail méditerranéen (*Corallium rubrum* L.) chez les peuples celtes: essai d'interprétation', in *Corallo di ieri, corallo di oggi*, ed. Jean-Paul Morel, Celia Rondi-Costanzo and Daniela Ugolini (Bari, 2000), pp. 193–203; Massimo Vidale et al., 'Symbols at War: The Impact of *Corallium rubrum* in the Indo-Pakistani Subcontinent', in *Ethnobiology of Corals and Coral Reefs*, ed. Nemer E. Narchi and Lisa L. Price (Cham, Heidelberg, New York, Dordrecht and London, 2015), p. 64.
11 Meeks, 'Le Corail dans l'Égypte ancienne'.
12 Tomoya Akimichi, 'Coral Trading and Tibetan Culture', in *A Biohistory of Precious Corals: Scientific, Cultural and Historical Perspectives*, ed. Nozomu Iwasaki (Kanagawa, 2010), pp. 149–62 (p. 154).
13 Pippa Lacey, 'The Coral Network: The Trade of Red Coral to the Qing Imperial Court in the Eighteenth Century', in *The Global Lives of Things: The Material Culture of Connections in the Early Modern World*, ed. Anne Gerritsen and Giorgio Riello (Abingdon, Oxon and New York, 2016), pp. 81–102 (p. 84).
14 Andrew Lawler, 'Sailing Sinbad's Seas', *Science*, CCCXLIV/6191 (2014), pp. 1440–45.
15 Xinru Liu, *The Silk Road in World History* (Oxford, 2010), p. 54.
16 Gedalia Yogev, *Diamonds and Coral: Anglo-Dutch Jews and Eighteenth-century Trade* (Leicester and New York, 1978), p. 103.
17 Akimichi, 'Coral Trading and Tibetan Culture', p. 155.
18 Vidale et al., 'Symbols at War'.
19 Roderich Ptak, 'Notes on the Word *Shanhu* and Chinese Coral Imports from Maritime Asia *c.* 1250–1600', *Archipel*, XXXIX/1 (1990), pp. 65–80 (p. 71).
20 Lacey, 'The Coral Network', p. 86.
21 Edrîsi, *Description de l'Afrique et de l'Espagne*, trans. R. Dozy and M. J. de Goeje (Leyden, 1866), p. 201.
22 François Doumenge, 'Le Corail rouge', in *Parures de la mer*, exh. cat., Musée Océanographique de Monaco (2000), p. 23.
23 Yvonne Hackenbroch, 'A Set of Knife, Fork, and Spoon with Coral Handles', *Metropolitan Museum Journal*, XV (1981), pp. 183–4.
24 Paula Gershick Ben-Amos, *Art, Innovation, and Politics in Eighteenth-century Benin* (Bloomington, IN, 1999), p. 83.
25 Sadao Kosuge, 'History of the Precious Coral Fisheries in Japan (1)', *Precious Corals and Octocoral Research*, 1 (1993), pp. 30–38.
26 Nishie, 'Precious Coral from a Cultural Perspective', p. 104.
27 Shinichiro Ogi, 'Coral Fishery and Kochi Prefecture in Modern Times', in *A Biohistory of Precious Corals: Scientific, Cultural and Historical Perspectives*, ed. Nozomu Iwasaki (Kanagawa, 2010), pp. 199–249 (p. 244).
28 Margaret Flower, *Victorian Jewellery* (London, 1951), p. 18.
29 James M. Cornelius, Curator, Abraham Lincoln Presidential Library and Museum, 'March 2012 Artifact of the Month: Mary Lincoln's Jewelry', www.youtube.com, accessed 22 October 2014.
30 Cristina Del Mare, 'Spanish Influence and the Introduction of Mediterranean Coral into America', in *The Coral Story: Coral, A Brief History of Mediterranean Coral*, www.traderoots.com, accessed 11 October 2014.
31 Margery Bedinger, *Indian Silver: Navajo and Pueblo Jewelers* (Albuquerque, NM, 1973), pp. 187–8.
32 Giovanni Tescione, *The Italians and their Coral Fishing*, trans. Maria Teresa Barke (Naples, 1968), p. 28.
33 Narcís Monturiol, *Ensayo sobre el arte de navegar por debajo del agua* (Valladolid, 2010 [1891]), pp. 36–7.
34 Matthew Stewart, *Monturiol's Dream: The Extraordinary Story of the Submarine Inventor who Wanted to Save the World* (New York, 2003), pp. 97–8.
35 Leonardo Fusco, *Red Gold*, trans. William Trubridge (Naples, 2011).
36 Ibid., p. 180.
37 Ibid., p. 246.
38 Kosuge, 'History of the Precious Coral Fisheries in Japan', p. 32.
39 Ogi, 'Coral Fishery and the Kuroshio Region in Modern Japan', p. 165.
40 Kosuge, 'History of the Precious Coral Fisheries in Japan', p. 33.

41 Les Watling, University of Hawaii, personal communication, email, 10 December 2014.
42 Doumenge, 'Le Corail rouge', pp. 22–3.
43 Basilio Liverino, *Red Coral: Jewel of the Sea*, trans. Jane Helen Johnson (Bologna, 1989), p. 74.
44 Doumenge, 'Le Corail rouge', pp. 23–4.
45 Olivier Lopez, 'Vivre et travailler pour la Compagnie royale d'Afrique en Barbarie au XVIIIe siècle', *Rives méditerranéennes*, 45, 'L'Histoire économique entre France et Espagne (XIXe–XXe siècles)' (2013), pp. 91–119, available at www.academia.edu/7503896, accessed 19 December 2014.
46 Caterina Ascione, 'The Art of Coral: Myth, History and Manufacture from Ancient Times to the Present', in *Il corallo rosso in Mediterraeo: arte, storia e scienzia / Red Coral in the Mediterranean Sea: Art, History and Science*, ed. F. Cicogna and R. Cattaneo-Vietti (Rome, 1993), pp. 25–36.
47 Liverino, *Red Coral*, p. 77.
48 Basilio Liverino, 'Fishing', www.liverino.it/english/pesca.htm, accessed 8 January 2015.
49 Georgios Tsunis et al., 'The Exploitation and Conservation of Precious Corals', *Oceanography and Marine Biology: An Annual Review*, XLVIII (2010), pp. 161–212.
50 Shinichiro Ogi, 'Coral Fishery and Kochi Prefecture in Modern Times', p. 243.
51 Nishie, 'Precious Coral from a Cultural Perspective', p. 93.
52 A. W. Bruckner, 'Advances in Management of Precious Corals in the Family Corallidae: Are New Measures Adequate?', *Current Opinion in Environmental Sustainability*, VII (2014), pp. 1–8.
53 Museu Marítim de Barcelona, *La Pesca del corall a Catalunya* (Barcelona, n.d.), p. 17.
54 Tsunis et al., 'The Exploitation and Conservation of Precious Corals', pp. 161–212.
55 Shui-Kai Chang, Ya-Ching Yang and Nozomu Iwasaki, 'Whether to Employ Trade Controls or Fisheries Management to Conserve Precious Corals (Corallidae) in the Northern Pacific Ocean', *Marine Policy*, XXXIX (2013), pp. 144–53.
56 Nozomu Iwasaki et al., 'Morphometry and Population Structure of Non-harvested and Harvested Populations of the Japanese Red Coral (*Paracorallium japonicum*) of Amami Island, Southern Japan', *Marine and Freshwater Research*, LXIII/5 (2012), pp. 468–74.
57 Andrew W. Bruckner, 'Advances in Management of Precious Corals to Address Unsustainable and Destructive Harvest Techniques', in *The Cnidaria, Past, Present and Future: The World of Medusa and her Sisters*, ed. Stefano Goffredo and Zvy Dubinsky (Cham, 2016), pp. 747–86.
58 Richard W. Grigg, *The Precious Corals: Fishery Management Plan of the Western Pacific Regional Fishery Management Council*, Pacific Islands Fishery Monographs, I (Honolulu, HI, 2010).
59 Richard W. Grigg, 'Precious Coral Fisheries of Hawaii and the U.S. Pacific Islands', *Marine Fisheries Review*, LV/2 (1993), pp. 50–60.
60 Chang, Yang and Iwasaki, 'Whether to Employ Trade Controls or Fishereries Management', p. 150.
61 Reiji Yoshida, 'Chinese coral poachers encroaching on Japanese fishermen', www.japantimes.co.jp, accessed 7 November 2017.
62 Campaign launched in 2008 by Seaweb and Tiffany & Co; see www.tiffanyandcofoundation.org.

6 Coral Construction

1 Anonymous, trans., *Voyage de M. Niebuhr en Arabie et en d'autres pays de l'Orient, avec l'extrait de la description de l'Arabie & des observations de Mr. Forskal* (Switzerland, 1780), p. 364.
2 A. L. Kroeber, *Anthropology*, revd edn (New York, 1948), pp. 255–6.
3 M. Elleray, 'Little Builders: Coral Insects, Missionary Culture and the Victorian Child', *Victorian Literature and Culture*, XXXIX/1 (2011), pp. 223–38 (p. 226).
4 Anonymous, 'Coral Rings', *Blackwood's Edinburgh Magazine*, LXXIV/455 (1853), pp. 360–71.
5 [Charles Dickens], review of 'The Poetry of Science, or Studies of the Physical Phenomena of Nature. By Robert Hunt', *The Examiner* (9 December 1848), p. 787.
6 Jules Rengade [Aristide Roger], *Voyage sous les flots*, 3rd edn (Paris, 1869), pp. 157, 159–60.
7 Robert Louis Stevenson, 'Edinburgh: Picturesque Notes, 1878', quoted and annotated in Rosalind Williams, *The Triumph of Human Empire: Verne, Morris and Stevenson at the End of the World* (Chicago, IL, 2013), pp. 239, 388.
8 Jules Michelet, 'Blood Flower', in *The Sea* [1861] (New York, 1864), pp. 144–5.
9 James Hamilton-Paterson, *Playing with Water* [1987] (New York, 1994), pp. 5, 8–9.
10 James Hamilton-Paterson, 'Reefs and Seeing', in *The Great Deep: The Sea and its Thresholds* (New York, 1992), p. 111.
11 *Georges Méliès: Encore – New Discoveries (1896–1911)*, DVD, Film Preservation Associates, Inc. and Flicker Alley LLC (Los Angeles, CA, 2010).
12 Vincent Callibaut, 'Coral Reef: Matrix and Plug-in for the Construction of 1,000 Passive Houses in Haiti', www.vincent.callebaut.org, accessed 27 February 2015.
13 Michail Vanis, 'Neo-nature Ch. 1: Animalia', www.mikevanis.com, accessed 5 May 2017.
14 Thomas J. Goreau and Wolf Hilbertz, 'Bottom-up Community-based Coral Reef and Fisheries Restoration in Indonesia, Panama, and Palau', in *Handbook of Regenerative Landscape Design*, ed. Robert L. France (Boca Raton, FL, 2008), pp. 143–60.
15 Joseph Banks's *Endeavour Journal*, quoted in Patrick O'Brian, *Joseph Banks: A Life* (Boston, MA, 1993), p. 101.

16 Warren D. Sharp et al., 'Rapid Evolution of Ritual Architecture in Central Polynesia Indicated by Precise 230Th/U Coral Dating', *Proceedings of the National Academy of Sciences USA*, CVI/30 (2010), pp. 13234–9.
17 Patrick V. Kirch, Regina Mertz-Kraus and Warren D. Sharp, 'Precise Chronology of Polynesian Temple Construction and Use for Southeastern Maui, Hawaiian Islands, Determined by ^{230}Th Dating of Corals', *Journal of Archaeological Science*, LIII (2015), pp. 166–77.
18 David Maxwell, 'Beyond Maritime Symbolism: Toxic Marine Objects from Ritual Contexts at Tikal', *Ancient Mesoamerica*, XI/1 (2000), pp. 91–8.
19 Heather McKillop et al., 'The Coral Foundations of Coastal Maya Architecture', in *Research Reports in Belizean Archaeology*, ed. Jaime Awe, John Morris and Sherilyne Jones (Belmopan, 2004), vol. I, pp. 347–58.
20 J. C. Andersen, *Myths and Legends of the Polynesians* (London, 1928), p. 456.
21 World Heritage Committee, *Taonga Pasifika: World Heritage in the Pacific* (Christchurch, 2007), pp. 12–13.
22 'Skull Island / Solomon Islands / Oceania', www.traveladventures.org, accessed 15 May 2014.
23 'Coral Construction', www.totakeresponsibility.blogspot.com, 13 October 2012.
24 Andrew Lawler, 'Sailing Sinbad's Seas', *Science*, CCCXLIV/6191 (2014), pp. 1440–45.
25 Dimitri Meeks, 'Le Corail dans l'Egypte ancienne', in *Corallo di ieri, corallo di oggi*, ed. Jean-Paul Morel, Celia Rondi-Costanzo and Daniela Ugolini (Bari, 2000), pp. 99–117 (pp. 110–11).
26 Michael Mallinson, personal communication in emails of 25 and 26 September 2017.
27 Jean-Pierre Greenlaw, *The Coral Buildings of Suakin: Islamic Architecture, Planning, Design and Domestic Arrangements in a Red Sea Port* (London and New York, 1995 [1976]); Jacke Phillips, 'Beit Khorshid Effendi: A Trader's House at Suakin', in *Navigated Spaces, Connected Places. Proceedings of Red Sea Project V. British Society for the Study of Arabia Monographs No. 12*, ed. Dionisius A. Agius, John P. Cooper, Athena Trakadas and Chiara Zazzaro (Oxford, 2012), pp. 187–99; Nancy Um, 'Reflections on the Red Sea Style: Beyond the Surface of Coastal Architecture', *Northeast African Studies*, 12 (2012), pp. 243–72.
28 Michael Mallinsom et al., chap. 24, 'Ottoman Suakin 1541–1865: Lost and Found', in *The Frontiers of the Ottoman World*, ed. A.C.S. Peacock (Oxford and New York, 2009), pp. 469–92; British Academy Scholarship Online: January 2012, DOI:10.5871/bacad/9780197264423.001.0001.
29 Phillips, 'Beit Khorshid Effendi'.
30 Barbara E. Brown, 'Mining/Quarrying of Coral Reefs', in *Encyclopedia of Modern Coral Reefs*, ed. David Hopley (Dordrecht, 2011), pp. 707–11.
31 Guillermo Horta-Puga, 'Environmental Impacts', in *Coral Reefs of the Southern Gulf of Mexico*, ed. J. W. Tunnell Jr, Ernesto A. Chávez and Kim Withers (College Station, TX, 2007), pp. 129–30.
32 Mike Dash, *Batavia's Graveyard* (New York, 2001), p. 250.
33 Hugh Edwards, *Islands of Angry Ghosts* (New York, 1966), pp. 54, 68–9.
34 'Coral Construction', www.totakeresponsibility.blogspot.com.
35 William F. Luce, 'Airfields in the Pacific', *Civil Engineering*, XV/10 (1945), pp. 453–4.
36 P. J. Halloran, 'Building B-29 Bases on Tinian Island', *Engineering News-Record*, CXXXV/9 (1945), pp. 302–7.
37 Nathan A. Bowers, 'Airfields of Coral', in W. G. Bowman et al., *Bulldozers Come First: The Story of U.S. War Construction in Foreign Lands* (New York and London, 1944), p. 176.
38 Vytautas B. Bandjunis, *Diego Garcia: Creation of the Indian Ocean Base* (Lincoln, NE, 2001), p. 55.
39 Ben Dolven et al., 'Chinese Land Reclamation in the South China Sea: Implications and Policy Options', *Congressional Research Service*, 25 pp., https://fas.org, 18 June 2005, accessed 1 July 2015.
40 Youna Lyons, quoted in Greg Torode, 'Paving Paradise: Scientists Alarmed over China Island Building in Disputed Sea', www.reuters.com, 25 June 2015.
41 John W. McManus, 'The Spratly Islands: A Marine Park?', *Ambio*, XXIII/3 (1994), pp. 181–6.
42 Walter M. Goldberg, 'Reefs Now and in the Next 100 Years', in *The Biology of Reefs and Reef Organisms* (Chicago, IL, and London, 2013), p. 340.
43 Peter W. Glynn and Mitchell Colgan, personal communication, email, 13 February 2017.
44 Rusty McClure and Jack Heffron, *Coral Castle: The Mystery of Ed Leedskalnin and his Coral Castle* (Dublin, OH, 2009).
45 Salvador Dalí, *The Secret Life of Salvador Dalí*, trans. Haakon M. Chevalier (Figueres, 1986), p. 218.

7 A New Age of Corals
1 Jean-Pierre Gattuso, Ove Hoegh-Guldberg and Hans-Otto Pörtner, 'Cross-chapter Box on Coral Reefs', in *Part A: Global and Sectoral Aspects. Contribution of Working Group II to the Fifth Assessment Report of the Intergovernmental Panel on Climate Change, Climate Change 2014: Impacts, Adaptation, and Vulnerability*, ed. C. B. Field et al. (Cambridge and New York, 2014), pp. 97–100.
2 Lauretta Burke et al., 'Executive Summary', in *Reefs at Risk Revisited* (Washington, DC, 2011), pp. 1–7.
3 Paul S. C. Taçon, Meredith Wilson and Christopher Chippindale, 'Birth of the Rainbow Serpent in Arnhem Land Rock Art and Oral History', *Archaeology of Oceania*, XXXI/3 (1996), pp. 103–24.

4 Zane Saunders, personal communication, email, 12 and 20 August 2015.
5 Stephen Jay Gould, 'Borderlines and Categories', in *Alexis Rockman*, with essays by Stephen Jay Gould, Jonathan Carey and David Quammen (New York, 2003), pp. 14–17.
6 Courtney Mattison, 'Sculpting the Beauty and Peril of Coral Reefs', *American Scientist*, CIII/4 (July–August 2015), pp. 292–6.
7 Margaret Wertheim and Christine Wertheim, *Crochet Coral Reef* (Los Angeles, CA, 2015).
8 Russ Van Arsdale, 'Fraudulent Huckster's Reward: 10 Easy Payments of One Year Each', http://bangordailynews.com, 23 March 2014.
9 *Kevin Trudeau v. Federal Trade Commission*, www.ftc.gov/system/files/documents/cases/120823_trudeaubrief.pdf, accessed 9 March 2015.
10 Stefan Helmreich, 'How Like a Reef: Figuring Coral, 1839–2010', in *Sounding the Limits of Life: Essays in the Anthropology of Biology and Beyond* (Princeton, NJ, 2016), pp. 48–61, p. 230 n. 36.
11 Lucas Leyva, interviewed in the film *Coral City*, part 1, dir. John McSwain (2014), http://creators.vice.com, accessed 17 June 2015.
12 J.-P. Gattuso et al., 'Contrasting Futures for Ocean and Society from Different CO_2 Emissions Scenarios', *Science*, CCCXLIX/6243 (2015), pp. 45–55.
13 Elizabeth Kolbert, 'The Siege of Miami', *New Yorker* (21 and 28 December 2015), pp. 42–6, 49–50.
14 Kenneth R. Weiss, 'Before We Drown We May Die of Thirst', *Nature*, DXXVI (2015), pp. 624–7.
15 Alexandre K. Magnan et al., 'Implications of the Paris Agreement for the Ocean', *Nature Climate Change*, VI/8 (2016), pp. 732–5.
16 Eli Kintisch, 'Climate Crossroads' and 'After Paris: The Rocky Road Ahead', *Science*, CCCL/6264 (2015), pp. 1017–19.
17 Willem Renema et al., 'Are Coral Reefs Victims of their Own Past Success?', *Science Advances*, II/4 (2016), DOI 10.1126/sciadv.1500850.
18 J.B.C. Jackson, 'Reefs since Columbus', *Coral Reefs*, XVI, suppl. 1 (1997), pp. S23–S32.
19 Nancy Knowlton and Jeremy B. C. Jackson, 'Shifting Baselines, Local Impacts, and Global Change on Coral Reefs', *PLoS Biology*, VI/2 (2008), p. e54.
20 Jennifer E. Smith et al., 'Re-evaluating the Health of Coral Reef Communities: Baselines and Evidence for Human Impacts across the Central Pacific', *Proceedings of the Royal Society B*, CCLXXXIII/1822 (2016), article 20151985, DOI 10.1098/rspb.2015.1985.
21 Jan Sapp, 'Crown of Thorns Inquisition', and 'Cassandra and the Sea Star', in *What is Natural? Coral Reef Crisis* (Oxford, 1999), pp. 65–76 and pp. 203–16 respectively.
22 Editorial, 'Ocean Preserves Necessary for Research, Recovery', *Bangor* [Maine] *Daily News*, 19 September 2016, p. A4.
23 Ginger Strand, 'Sea Change', *Nature Conservancy Magazine* (December 2016–January 2017), pp. 30–41.
24 Paul L. Jokiel, 'Temperature Stress and Coral Bleaching', in *Coral Health and Disease*, ed. Eugene Rosenberg and Yossi Loya (Berlin, Heidelberg and New York, 2004), pp. 401–25.
25 Y. Loya et al., 'Coral Bleaching: The Winners and the Losers', *Ecology Letters*, IV/2 (2001), pp. 122–31.
26 P. W. Glynn, 'Widespread Coral Mortality and the 1982–83 El Niño Warming Event', *Environmental Conservation*, XI/2 (1984), pp. 133–46; P. W. Glynn and L. D'Croz, 'Experimental Evidence for High Temperature Stress as the Cause of El Niño-Coincident Coral Mortality', *Coral Reefs*, VIII/4 (1990), pp. 181–91.
27 J. K. Oliver, R. Berkelmans and C. M. Eakin, 'Coral Bleaching in Space and Time', in *Coral Bleaching: Patterns, Processes, Causes and Consequences*, ed. Madeleine J. H. van Oppen and Janice M. Lough (Berlin and Heidelberg, 2009), pp. 21–39.
28 Thomas J. Goreau and Raymond L. Hayes, 'Coral Bleaching and Ocean "Hot Spots"', *Ambio*, XXIII/3 (1994), pp. 176–80.
29 Jokiel, 'Temperature Stress and Coral Bleaching', Table 23.1.
30 Simon D. Donner et al., 'Global Assessment of Coral Bleaching and Required Rates of Adaptation under Climate Change', *Global Change Biology*, XI/12 (2005), pp. 2251–65.
31 Terry P. Hughes et (45) al., 'Global Warming and Recurrent Mass Bleaching of Corals', *Nature*, DXLIII (2017), pp. 373–7.
32 Michael P. Lesser, 'Oxidative Stress in Marine Environments: Biochemistry and Physiological Ecology', *Annual Review of Physiology*, LXVIII (2006), pp. 253–78.
33 Climate Scoreboard, www.climateinteractive.org/programs/scoreboard, accessed 27 April 2016.
34 NOAA, 'El Niño Prolongs Longest Global Coral Bleaching Event', www.noaa.gov, 23 February 2016.
35 Tom Arup, 'Startling Images Reveal Devastating Bleaching on the Great Barrier Reef', *Sydney Morning Herald*, 22 March 2016, www.smh.com.au.
36 Christopher Pala, 'Corals Tie Stronger El Niños to Climate Change', *Science*, CCCLIV/6317 (2016), p. 1210.
37 Nicky Phillips, 'Australian Election Gives Climate Researchers Hope', *Nature*, DXXXIV (2016), pp. 433–4.
38 Katherine Gregory, 'Barrier Reef at Risk of Winding Up on UNESCO "Danger" List, Queensland Government Says', www.abc.net.au/news, 1 December 2016.
39 Editorial, 'Scientists Must Fight for the Facts', *Nature*, DXLI/7638 (2017), p. 435.
40 J.E.N. Veron, 'Messages from Deep Time', in *A Reef in Time: The Great Barrier Reef from Beginning to End* (Cambridge, MA, and London), p. 98.

41 Wolfgang Kiessling, 'Geologic and Biologic Controls on the Evolution of Reefs', *Annual Review of Ecology, Evolution, and Systematics*, XL (2009), pp. 173–92 (p. 185).
42 'Carbon Dioxide in Earth's Atmosphere', https://en.wikipedia.org, accessed 7 April 2017.
43 Kintisch, 'After Paris', p. 1018.
44 Jeff Goodell, 'Will Paris Save the World?', *Rolling Stone* (28 January 2016), pp. 28–33.
45 Alexander C. Gagnon, 'Coral Calcification Feels the Acid', *Proceedings of the National Academy of Sciences USA*, CX/5 (2013), pp. 1567–8; Justin Ries, 'Acid Ocean Cover Up', *Nature Climate Change*, 1 (2011), pp. 294–5.
46 Ove Hoegh-Guldberg, 'Coral Reef Sustainability through Adaptation: Glimmer of Hope or Persistent Mirage?', *Current Opinion in Environmental Sustainability*, VII (2014), pp. 127–33.
47 M. O. Clarkson et al., 'Ocean Acidification and the Permo-Triassic Mass Extinction', *Science*, CCCXLVIII/6231 (2015), pp. 229–32.
48 International Geosphere–Biosphere Programme, Intergovernmental Oceanographic Commission, and Scientific Committee on Oceanic Research, *Ocean Acidification Summary for Policymakers – Third Symposium on the Ocean in a High-CO_2 World* (Stockholm, 2013), p. 16.
49 Papua New Guinea: Katharina E. Fabricius et al., 'Losers and Winners in Coral Reefs Acclimatized to Elevated Carbon Dioxide Concentrations', *Nature Climate Change*, 1 (2011), pp. 165–9; Caroline Islands: I. C. Enochs et al., 'Shift from Coral to Macroalgae Dominance on a Volcanically Acidified Reef', *Nature Climate Change*, 5 (2015), pp. 1083–8; Ryukyu Islands: Shihori Inoue et al., 'Spatial Community Shift from Hard to Soft Corals in Acidified Water', *Nature Climate Change*, 3 (2013), pp. 683–7; Yucatán: Elizabeth D. Crook et al., 'Reduced Calcification and Lack of Acclimatization by Coral Colonies Growing in Areas of Persistent Natural Acidification', *Proceedings of the National Academy of Sciences USA*, CX/27 (2013), pp. 11044–9.
50 Kathryn E. F. Shamberger et al., 'Diverse Coral Communities in Naturally Acidified Waters of a Western Pacific Reef', *Geophysical Research Letters*, XLI/2 (2014), pp. 499–504.
51 Summarized by Janice M. Lough, 'Turning Back Time', *Nature*, DXXXI (2016), pp. 314–15.
52 J. Silverman et al., 'Community Calcification in Lizard Island, Great Barrier Reef: A 33-year Perspective', *Geochimica et Cosmochimica Acta*, CXLIV (2014), pp. 72–81.
53 Joan A. Kleypas and Chris Langdon, 'Coral Reefs and Changing Seawater Carbonate Chemistry', in *Coral Reefs and Climate Change: Science and Management*, ed. Jonathan T. Phinney et al. (Washington, DC, 2006), pp. 73–110 (p. 77).
54 Andreas J. Andersson, Fred T. Mackenzie and Jean-Pierre Gattuso, 'Effects of Ocean Acidification on Benthic Processes, Organisms, and Ecosystems', in *Ocean Acidification*, ed. Jean-Pierre Gattuso and Lina Hansson (Oxford, 2011), pp. 122–53 (p. 141 fig 7.2).
55 Joy N. Smith et al., 'Ocean Acidification Reduces Demersal Zooplankton That Reside in Tropical Coral Reefs', *Nature Climate Change*, 6 (2016), pp. 1124–30.
56 Maoz Fine and Dan Tchernov, 'Scleractinian Coral Species Survive and Recover from Decalcification', *Science*, CCCXV/5820 (2007), p. 1811.
57 Mónica Medina et al., 'Naked Corals: Skeleton Loss in Scleractinia', *Proceedings of the National Academy of Sciences USA*, CIII/24 (2006), pp. 9096–100.
58 Cornelia Maier et al., 'End of the Century CO_2 Levels Do Not Impact Calcification in Mediterranean Cold-water Corals', *PLoS One*, VIII/4 (2013), p. e62655.
59 Les Watling, 'Deep-sea Trawling Must Be Banned', *Nature*, DI (2013), p. 7.
60 G. Roff, 'Earliest Record of a Coral Disease from the Indo-Pacific?', *Coral Reefs*, XXXV/2 (2016), p. 457.
61 Drew Harvell et al., 'Coral Disease, Environmental Drivers, and the Balance between Coral and Microbial Associates', *Oceanography*, XX/1 (2007), pp. 172–95.
62 Kathryn L. Patterson et al., 'The Etiology of White Pox, a Lethal Disease of the Caribbean Elkhorn Coral, *Acropora palmata*', *Proceedings of the National Academy of Sciences USA*, XCIX/13 (2002), pp. 8725–30.
63 Forest Rohwer with Merry Youle, 'Coral Diseases', in *Coral Reefs in the Microbial Seas* (Basalt, CO, 2010), pp. 69–84 (p. 75); Michael Thrusfield, 'Etiology', in *Diseases of Coral*, ed. Cheryl M. Woodley et al. (Hoboken, NJ, 2016), pp. 6–27.
64 Rohwer and Youle, 'Coral Diseases', pp. 76–7.
65 B.C.C. Hume et al., '*Symbiodinium thermophilum* sp. nov., a Thermotolerant Symbiotic Alga Prevalent in Corals of the World's Hottest Sea, the Persian/Arabian Gulf', *Scientific Reports*, 5 (2015), article 8562; DOI 10.1038/srep08562.
66 Yimnong Golbuu et al., 'Long-term Isolation and Local Adaptation in Palau's Nikko Bay Help Corals Thrive in Acidic Waters', *Coral Reefs*, XXXV/3 (2016), pp. 909–19, DOI 10.1007/s00338-016-1457-5.
67 Tracy D. Ainsworth et al., 'Climate Change Disables Coral Bleaching Protection on the Great Barrier Reef', *Science*, CCCLII/6283 (2016), pp. 338–42.
68 Ibid., p. 340.
69 Ove Hoegh-Guldberg, 'The Adaptation of Coral Reefs to Climate Change: Is the Red Queen Being Outpaced?', *Scientia Marina*, LXXVI/2 (2012), pp. 403–8.
70 Joan Kleypas, 'Invisible Barriers to Dispersal', *Science*, CCCXLVIII/6239 (2015), pp. 1086–7; Paul R. Muir et al., 'Limited Scope for Latitudinal Extension of Reef Corals', *Science*, CCCXLVIII/6239 (2015), pp. 1135–8.
71 S. Watanabe et al., 'Future Projections of Surface UV-B in a Changing Climate', *Journal of Geophysical Research*, CXVI/D16 (2011), DOI 10.1029/2011jd015749.

72 Y. Yara et al., 'Ocean Acidification Limits Temperature-induced Poleward Expansion of Coral Habitats around Japan', *Biogeosciences*, IX (2012), pp. 4955–68.
73 Groves B. Dixon et al., 'Genomic Determinants of Coral Heat Tolerance across Latitudes', *Science*, CCCXLVIII/6242 (2015), pp. 1460–62.
74 Stephen R. Palumbi et al., 'Mechanisms of Reef Coral Resistance to Future Climate Change', *Science*, CCCXLIV/6186 (2014), pp. 895–8.
75 C. Mark Eakin, 'Lamarck Was Partially Right – and That Is Good for Corals', *Science*, CCCXLIV/6186 (2014), pp. 798–9.
76 Maren Ziegler et al., 'Bacterial Community Dynamics Are Linked to Patterns of Coral Heat Tolerance', *Nature Communications*, 8:14213 (2017) [DOI:10.1038/ncomms14213]; Gergely Torda et al., 'Rapid Adaptive Responses to Climate Change in Corals', *Nature Climate Change*, 7 (2017), pp. 627–36.
77 Madeleine J. H. van Oppen et al., 'Novel Genetic Diversity through Somatic Mutations: Fuel for Adaptation of Reef Corals?', *Diversity*, III/3 (2011), pp. 405–23.
78 Maximilian Schweinsberg et al., 'More Than One Genotype: How Common is Intracolonial Genetic Variability in Scleractinian Corals?', *Molecular Ecology*, XXIV/11 (2015), pp. 2673–85.
79 Baruch Rinkevich et al., 'Venturing in Coral Larval Chimerism: A Compact Functional Domain with Fostered Genotypic Diversity', *Scientific Reports*, 6 (2016), DOI 10.1038/srep19493.
80 Cheryl A. Logan et al., 'Incorporating Adaptive Responses into Future Projections of Coral Bleaching', *Global Change Biology*, XX/1 (2014), pp. 125–39, DOI 10.1111/gcb.12390.
81 'Ecosystem Services', http://oceanwealth.org/ecosystem-services, accessed 4 January 2017.
82 Madeleine J. H. van Oppen et al., 'Building Coral Reef Resilience through Assisted Evolution', *Proceedings of the National Academy of Sciences USA*, CXII/8 (2015), pp. 2307–13.
83 Ken Caldeira, quoted in Elizabeth Kolbert, 'Unnatural Selection', *New Yorker* (18 April 2016), pp. 22–8 (p. 26).
84 Ruth Gates, quoted in Kolbert, 'Unnatural Selection', p. 28.
85 Thomas J. Goreau and Robert Kent Trench, eds, *Innovative Methods of Marine Ecosystem Restoration* (Boca Raton, FL, 2013).
86 Baruch Rinkevich, 'Rebuilding Coral Reefs: Does Active Reef Restoration Lead to Sustainable Reefs?', *Current Opinion in Environmental Sustainability*, VII (2014), pp. 28–36.
87 Ibid., p. 33.
88 Amanda Mascarelli, 'Designer Reefs', *Nature*, DVIII (2014), pp. 444–6.
89 Dixon et al., 'Genomic Determinants of Coral Heat Tolerance'.
90 Stephen R. Palumbi and Anthony R. Palumbi, *Extreme Life of the Sea* (Princeton, NJ, and Oxford, 2014), p. 160.
91 'Global Coral Bleaching – 2015/2016. The World's Third Major Global Event Now Confirmed', www.globalcoralbleaching.org, accessed 10 October 2015.
92 Dennis Normile, 'El Niño's Warmth Devastating Reefs Worldwide', *Science*, CCCLII/6281 (2016), pp. 15–16.
93 Hughes et al., 'Global Warming and Recurrent Mass Bleaching of Corals'.
94 Great Barrier Reef Marine Park Authority, 'Second Wave of Mass Bleaching Unfolding on Great Barrier Reef', www.gbrmpa.gov.au, 10 March 2017.
95 'IEA finds CO_2 emissions flat for third straight year even as global economy grew in 2016', 17 March 2017, www.iea.org/newsroom/news/2017/march/iea-finds-co2-emissions-flat-for-third-straight-year-even-as-global-economy-grew.html, accessed 13 October 2017.
96 Ibid.
97 C. Mark Eakin et al., 'Ding, Dong, the Witch is Dead (?) – Three Years of Global Coral Bleaching 2014–2017', *Reef Encounter*, 32 (2017), pp. 33–8.

Coda: What Lies Ahead?
1 Judith Wright, *The Coral Battleground* [1977] (North Melbourne, 2014), p. 186.
2 For a recent broad consideration, see The Royal Society, *People and the Planet* (London, 2012).
3 Pope Francis, 'On Care for Our Common Home', http://w2.vatican.va/content/vatican/en.html, 24 May 2015.
4 Anthony J. McMichael, 'Introduction', in *Climate Change and the Health of Nations: Famines, Fevers, and the Fate of Populations* (Oxford, 2017), p. 21.
5 J.-P. Gattuso et (21) al., 'Contrasting Futures for Ocean and Society from Different CO_2 Emissions Scenarios', *Science*, CCCXLIX/6243 (2015), pp. 45–55.
6 Joeri Rogelj, Michel den Elzen, Niklas Höhne et al., 'Paris Agreement Climate Proposals Need a Boost to Keep Warming Well below 2°C', *Nature*, DXXXIV (2016), pp. 631–8.
7 Alexandre K. Magnan, Michel Colombier, Raphaël Billé et al., 'Implications of the Paris Agreement for the Ocean', *Nature Climate Change*, VI (2016), pp. 732–5.
8 Charlie Veron, *A Reef in Time* (Cambridge, MA, 2008), p. 231.
9 James Hamilton-Paterson, *Playing with Water* [1987] (New York, 1994), pp. 107, 155, 248.
10 Rowan Jacobsen, *A Geography of Oysters* (New York, 2008), p. 58.
11 'Ecosystem Services', http://oceanwealth.org/ecosystem-services, accessed 4 January 2017.
12 Richard Holmes, 'Travelling', in *This Long Pursuit: Reflections of a Romantic Biographer* (New York, 2016), p. 14.
13 Nancy Knowlton, 'Doom and Gloom Won't Save the World', *Nature*, 544 (2017), p. 271.

Select Bibliography

Birkeland, Charles, ed., *Coral Reefs in the Anthropocene* (Dordrecht, Heidelberg, New York and London, 2015)
—, ed., *Life and Death of Coral Reefs* (New York, 1997)
Bowen, James, *The Coral Reef Era: From Discovery to Decline* (Dordrecht, 2015)
—, and Margarita Bowen, *The Great Barrier Reef: History, Science, Heritage* (Cambridge, 2002)
Catala, René L. A., *Carnival under the Sea* (Paris, 1964)
Dana, James D., *Corals and Coral Islands* (New York, 1872)
Darwin, Charles, *The Structure and Distribution of Coral Reefs* (London, 1842), http://darwin-online.org.uk, accessed 19 October 2015
Davidson, Osha Gray, *The Enchanted Braid: Coming to Terms with Nature on the Coral Reef* (New York, 1998)
Dobbs, David, *Reef Madness: Charles Darwin, Alexander Agassiz, and the Meaning of Coral* (New York, 2005)
Dubinsky, Zvy, and Noga Stambler, eds, *Coral Reefs: Ecosystems in Transition* (Dordrecht, Heidelberg, London and New York, 2011)
Ellis, John, *The Natural History of Many Curious and Uncommon Zoophytes* (London, 1786), at www.biodiversitylibrary.org, accessed 19 October 2015
Endt-Jones, Marion, ed., *Coral: Something Rich and Strange* (Liverpool, 2015)
Goffredo, Stefano, and Zvy Dubinsky, eds, *The Cnidaria, Past, Present and Future: The World of Medusa and Her Sisters* (Cham, 2016)
Goldberg, Walter M., *The Biology of Reefs and Reef Organisms* (Chicago, IL, 2013)
Goreau, Thomas J., and Robert Kent Trench, eds, *Innovative Methods of Marine Ecosystem Restoration* (Boca Raton, FL, 2013)
Gosse, Philip Henry, *Actinologia Britannica: A History of the British Sea-anemones and Corals* (London, 1860), at www.biodiversitylibrary.org, accessed 19 October 2015
Hopley, David, ed., *Encyclopedia of Modern Coral Reefs: Structure, Form and Process* (Dordrecht, 2011)
Iwasaki, Nozomu, ed., *A Biohistory of Precious Corals: Scientific, Cultural and Historical Perspectives* (Kanagawa, 2010)

Johnson, Johanna E., and Paul A. Marshall, eds, *Climate Change and the Great Barrier Reef: A Vulnerability Assessment* (Townsville, QLD, 2007)
Johnston, George, *A History of the British Zoophytes* (Edinburgh, London and Dublin, 1838)
Jones, Steve, *Coral: A Pessimist in Paradise* (London, 2007)
Liverino, Basilio, *Red Coral: Jewel of the Sea*, trans. Jane Helen Johnson (Bologna, 1989)
McCalman, Iain, *The Reef: A Passionate History* (New York, 2013)
Marsigli, Luigi Ferdinando, *Natural History of the Sea* [1725], trans. Anita McConnell (Bologna, 1999)
Narchi, Nemer E., and Lisa L. Price, eds, *Ethnobiology of Corals and Coral Reefs* (Cham, Heidelberg, New York, Dordrecht and London, 2015)
Riegl, Bernhard M., and Richard E. Dodge, eds, *Coral Reefs of the World*, vol. 1: *Coral Reefs of the USA* (Dordrecht and London, 2008)
Roberts, J. Murray, Andrew Wheeler, André Freiwald and Stephen Cairns, *Cold-water Corals: The Biology and Geology of Deep-sea Coral Habitats* (Cambridge, 2009)
Rohwer, Forest, with Merry Youle, *Coral Reefs in the Microbial Seas* (Basalt, CO, 2010)
Rosenberg, Eugene, and Yossi Loya, eds, *Coral Health and Disease* (Berlin, 2004)
Sapp, Jan, *What is Natural? Coral Reef Crisis* (Oxford, 1999)
Saville-Kent, William, *The Great Barrier Reef of Australia: Its Products and Potentialities* (London, 1893), at www.biodiversitylibrary.org, accessed 19 October 2015
Sheppard, Anne, *Coral Reefs: Secret Cities of the Sea* (London, 2015)
Sheppard, Charles, *Coral Reefs: A Very Short Introduction* (Oxford, 2014)
Sheppard, Charles R. C., Graham M. Pilling and Simon K. Davy, *The Biology of Coral Reefs* [2009] (Oxford, 2012)
Spalding, Mark D., Corinna Ravilious and Edmund P. Green, *World Atlas of Coral Reefs* (Berkeley, CA, 2001)
Sprung, Julian, and Charles Delbeek, *The Reef Aquarium*, 3 vols (Miami Gardens, FL, 1994, 1997, 2005)

Tescione, Giovanni, *The Italians and their Coral Fishing*, trans. Maria Teresa Barke (Naples, 1968)

Tunnell, John W., Jr, Ernesto A. Chávez and Kim Withers, eds, *Coral Reefs of the Southern Gulf of Mexico* (College Station, TX, 2007)

van Oppen, Madeleine J. H., and Janice M. Lough, eds, *Coral Bleaching: Patterns, Processes, Causes and Consequences* (Berlin, 2009)

Veron, J.E.N., *Corals in Space and Time: The Biogeography and Evolution of the Scleractinia* (Sydney, 1995)

—, *Corals of the World*, 3 vols (Townsville, QLD, 2000)

—, *A Reef in Time: The Great Barrier Reef from Beginning to End* (Cambridge, MA, 2008)

Wertheim, Margaret, and Christine Wertheim, *Crochet Coral Reef* (Los Angeles, CA, 2015)

Woodley, Cheryl M., Craig A. Downs, Andrew W. Bruckner, James W. Porter and Sylvia B. Galloway, eds, *Diseases of Coral* (Hoboken, NJ, 2016)

Wright, Judith, *The Coral Battleground*, 3rd edn [1977] (North Melbourne, VIC, 2014)

Yonge, C. M., *A Year on the Great Barrier Reef: The Story of Corals and the Greatest of their Creations* (London, 1930)

Glossary

acclimatization An organism's physiological or biochemical adjustment to compensate for an environmental change within its normal tolerance range, such as a seasonal change in temperature or daily cycles of solar radiation. Acclimatization does not involve a genetic change in the organism, but the limits to acclimatization are a PHENOTYPIC character determined by the genotype. By contrast, *see* ADAPTATION (process).

Actiniaria Order in the subclass HEXACORALLIA comprising the uncalcified sea anemones.

adaptation Heritable, genetically based character that contributes to an individual's fitness (survival and reproduction) in a particular environment. Adaptation also refers to the process of evolutionary (genetic) change in a population in response to a changed environment acting through natural selection of an adaptive PHENOTYPIC trait, usually requiring many generations.

Alcyonacea Order in the subclass OCTOCORALLIA comprising diverse 'soft corals'. Families of Alcyonacea are distinguished largely by the growth form of their colonies, the presence or absence of a supporting central axis, and the chemical nature (mineral or protein) of that axis. Abundant leathery alcyonaceans in the family Alcyoniidae may compete with SCLERACTINIAN hard corals for space on coral reefs.

Anthozoa (anthozoans) Class in the phylum CNIDARIA whose solitary or colonial members exist only as POLYPS, never as medusae. These 'flower animals' may lack a solid skeleton, or have a calcified or proteinaceous supporting framework. The two subclasses of the Anthozoa are HEXACORALLIA and OCTOCORALLIA.

Anthrobscene To emphasize the negative moral connotation of excessive exploitation of earth's resources and its environmental, especially toxic, consequences. The cultural theorist Jussi Parikka coined the neologism.

Anthropocene The geological epoch in which we live, when human activity has become the dominant agent changing the earth and its ecosystems; for example atmospheric composition, global climate change, deforestation, mining and species extinctions.

antioxidant Substance or enzyme in an organism that removes oxidizing agents such as the highly reactive hydroxyl or superoxide radicals, peroxides or other REACTIVE OXYGEN SPECIES (ROS) that are damaging to biomolecules in living organisms.

Antipatharia (antipatharians) Order in the subclass HEXACORALLIA comprising the black corals, uncalcified colonial POLYPS having a characteristic thorny, proteinaceous skeleton of antipathin, which takes a high polish and is used in making jewellery. The ages of some live specimens of black corals have been determined at more than four thousand years old, making them the oldest living animals.

aragonite (*adj.* **aragonitic**) Crystalline form of calcium carbonate in which the elongated crystal has three unequal axes at right angles to each other. Aragonite is secreted by SCLERACTINIAN HEXACORALS, the primary builders of coral reefs. Aragonite and CALCITE contain the same chemical elements but have different crystal structures.

asexual reproduction Vegetative growth or proliferation in which a coral colony increases its size by the budding or binary fission of a POLYP into two or more new polyps, all genetically identical to the parent; also, the generation of offspring from a single parent without the fusion of gametes but including the development of an unfertilized ovum. The symbiotic (*see* symbiosis) DINOFLAGELLATES (ZOOXANTHELLAE) living inside the coral host's cells exist in a vegetative stage and by binary fission match their proliferation to that of the host's tissues.

atoll Ring-shaped, offshore (oceanic) coral reef or coral island consisting of a continuous or broken rim enclosing a central LAGOON. Atolls frequently occur as groups of islands, called archipelagos.

azooxanthellate coral Coral (most commonly, a SCLERACTINIAN) that does not naturally harbour symbiotic DINOFLAGELLATES called ZOOXANTHELLAE. Distinguished from zooxanthellate corals that normally form such symbioses.

bioherm Loosely consolidated reef (not limited to one made by corals) that lacks infilling and internal cementation, leaving it less massive and more porous.

bleaching *see* CORAL BLEACHING

bryozoan Member of the phylum **Bryozoa** (or Ectoprocta), called moss animals or ectoprocts, in which almost all species form colonies of many polyp-like individuals called zooids. Bryozoans were once classified among the taxonomically ambiguous ZOOPHYTES.

calcareous Characterized by having a skeleton composed of some mineral of calcium carbonate ($CaCO_3$).

calcite (*adj.* **calcitic**) Crystalline form of calcium carbonate consisting of regular hexagonal facets secreted by all calcareous taxa of OCTOCORALLIA, and by extinct corals (Rugosa and Tabulata). Magnesium may be substituted for some of the calcium, forming magnesium calcite.

Ceriantharia (**cerianthids**) Order in subclass HEXACORALLIA comprising the uncalcified tube anemones or cerianthids, which may be the most basal (earliest diverging) taxon of hexacorals. Unique to the cerianthids is the ptychocyst, a type of CNIDA used to build the tube.

chimera Organism (including a colony) that includes genetically distinct cells and tissues. Chimeric corals arise from fusion of different embryos or colonies.

clade Group or lineage of species consisting of the ancestral species and all its descendants.

clone (*adj.* **clonal**) Group of genetically identical organisms produced via some form of ASEXUAL REPRODUCTION; colloquially, an individual that is genetically identical to another.

cnida (*pl.* **cnidae**) Complex capsular organelle secreted by CNIDOCYTES (cells) and used for adhesion (spirocysts), stinging and capture of prey (NEMATOCYSTS), or tube building (ptychocysts) in various members of CNIDARIA, the phylum named for the unique organelle.

Cnidaria (**cnidarians**) Ancient phylum of multicellular animals defined by the presence of CNIDAE and which includes several TAXA commonly referred to as 'corals'.

coenenchyme Communal mass of MESOGLEA plus EPIDERMAL and GASTRODERMAL tissues that connects colonial POLYPS and constitutes most of the living mass of the COLONY, especially in octocorallian 'soft corals'.

coenosarc In colonial SCLERACTINIANS, the layer of tissues lying outside of the polyp's CORALLITE and connecting each polyp to the rest of the COLONY.

colonial Physically interconnected and physiologically integrated coral POLYPS formed by ASEXUAL proliferation of a founding polyp and successive polyps and supported by a common skeleton or POLYPARY of varying chemical composition (CALCAREOUS or proteinaceous).

coral bleaching Loss of all or some ZOOXANTHELLAE or their photosynthetic pigments from the coral host as a general response to environmental stress. Consequently the white calcium carbonate skeleton becomes visible through the translucent animal tissues, giving the colony an overall white, bleached appearance. **Mass coral bleaching** refers to a prolonged, geographically widespread bleaching event on many reefs as a response to persistent, anomalously high seawater temperatures and high solar irradiance.

Coral Triangle Region in the eastern Indian and southwestern Pacific oceans having nearly six hundred species of SCLERACTINIAN corals, perhaps 75 per cent of all scleractinian species worldwide; the triangle includes Indonesia, most of Malaysia and the Philippines, Papua New Guinea and the Solomon Islands.

Coralliidae Family of OCTOCORALLIA in the order ALCYONACEA that includes the so-called precious corals having red or pink (or rarely, white) CALCITIC central axes used in making jewellery and art objects.

Corallimorpharia (**corallimorpharians**) Order of solitary or colonial, uncalcified HEXACORALS resembling sea anemones but taxonomically closer to SCLERACTINIANS.

corallite Cup-like CALCAREOUS structure secreted by and surrounding an individual SCLERACTINIAN polyp, subdivided by SEPTA and part of the larger skeleton (**corallum**) of the entire colony.

cryptic species Two or more unrecognized separate SPECIES classified together under a single species name.

diagenesis Combined physical, chemical and biological processes that convert discrete sediments to sedimentary rock. In coral reefs, the cement involved includes crystals that form beneath the surface of the sedimentary debris, binding it together and reducing the porosity of the reef and enhancing its solidity.

dinoflagellates Phylum of unicellular algae, photosynthetic or heterotrophic. They can be free-living and motile as phytoplankton and move using two flagella; in their vegetative, asexual phase photosynthetic dinoflagellates form symbioses with many marine invertebrates. *See also* ZOOXANTHELLAE.

El Niño Periodic but not entirely predictable intrusion of warm seawater into the tropical eastern Pacific, driven by higher atmospheric pressure at the surface in the western Pacific, as the warm phase of the El Niño–Southern Oscillation (ENSO). ENSO-related elevated sea surface temperatures and other perturbations contribute to geographic mass bleaching of corals not only in the tropical eastern Pacific but also the central and western Pacific.

endosymbiosis SYMBIOSIS with one partner living inside the body of the other. In corals, endosymbiotic algae such as DINOFLAGELLATES (ZOOXANTHELLAE) commonly live inside GASTRODERMAL cells of the host.

epiderm(is) Outer layer of tissue in the body wall of a CNIDARIAN. Especially in the prey-capturing tentacles, this layer is rich in CNIDOCYTES that contain NEMATOCYSTS.

epigenetics Study of PHENOTYPIC changes in an organism brought about by alterations in GENE activity or EXPRESSION without changes in nucleic acids of the genes themselves, for example by molecules binding to the chromosomes. Some epigenetic changes may be heritable.

eugenics Historically, the study or doctrine of improving a population or species (originally directed towards humans) by selective breeding of individuals to increase the incidence of heritable traits considered to be beneficial or desirable, or to decrease the frequency of undesirable traits, in the population.

exoskeleton External skeleton, secreted by an animal such as a coral polyp, that surrounds and supports the animal and that contributes to the hard structure of the larger COLONY or POLYPARY in colonial species.

fluorescence Emission of light at a visible wavelength longer and of lower energy than that of the absorbed light (including invisible ultraviolet). Vivid fluorescent colours of many ANTHOZOANS are due to the presence in the animal host's tissue of diverse chromoproteins (for example, green fluorescent protein, GFP) whose differences in chemical structure account for the different wavelengths (colours) they emit.

gastroderm(is) (*adj.* **gastrodermal**) Inner layer of tissue in the body wall of a CNIDARIAN. ENDOSYMBIOTIC algae such as ZOOXANTHELLAE commonly occur inside gastrodermal cells.

gene expression Manifestation of the information encoded in a gene as synthesis of a gene product, for example, a protein involved in producing the PHENOTYPIC character associated with the gene.

hermaphrodite (*adj.* **hermaphroditic**) Individual having both male and female gonads.

Hexacorallia Subclass of ANTHOZOA including polyps whose central cavity is divided longitudinally by paired fleshy MESENTERIES that occur in multiples of six, arranged radially around the inner face of the body wall. Hexacorals include hard, reef-building SCLERACTINIAN corals as well as their close relatives the CORALLIMORPHARIANS, and the sea anemones (ACTINIARIA), tube anemones (CERIANTHIDS), black corals (ANTIPATHARIANS) and gold corals (ZOANTHIDS). *Compare* OCTOCORALLIA.

holobiont In restricted usage, the SYMBIOTIC partnership that includes both the animal host and its algal ENDOSYMBIONTS, but more widely, the host and all of its microbial associates, including algae, bacteria, fungi and viruses.

Hydrozoa (**hydrozoans**) One of three original classes of phylum CNIDARIA, now included in the subphylum MEDUSOZOA along with class SCYPHOZOA and other classes of medusoid cnidarians formerly classified as scyphozoans. The calcareous lace corals and the fire corals are hydrozoans.

indeterminate (colony) growth In corals and other cnidarians that show vegetative ASEXUAL PROLIFERATION, there is no fixed body size. Growth continues as a potentially unlimited collective accumulation of biomass in modular individuals that make up the colony.

intelligent design Creationist argument, not amenable to scientific verification, that posits that organisms are so complex that they must have been designed and created by an intelligent entity.

lagoon Fully or partly enclosed, shallow body of water bounded entirely (as an annular ATOLL) by the rims of coral islands, or by the shore of a larger landmass and an elongate offshore barrier reef.

Lithophyta (lithophytes) (mostly *arch.*) In Linnaeus's taxonomy, Lithophyta was one of the major TAXA containing corals – a mélange of hydrozoan and anthozoan corals (the latter including scleractinians and octocorals), united by being both calcified and colonial. Other naturalists such as Luigi Ferdinando Marsigli include among the lithophytes less calcified sea whips and other octocorals, which he considered to be woody marine plants.

madrepore (*arch.*) Member of the hard or stony corals, most of which were originally placed in the genus *Madrepora* (since divided into many genera) and included by Linnaeus as one taxon within LITHOPHYTA.

mass coral bleaching *see* CORAL BLEACHING

Medusozoa (medusozoans) Clade or subphylum of phylum CNIDARIA. Most medusozoan life cycles include a medusa (most often, a mobile 'jellyfish'). Within the Medusozoa, only the class HYDROZOA includes some forms referred to as 'corals'.

mesentery Internal projection of the body wall of an ANTHOZOAN POLYP comprising the MESOGLEA and GASTRODERMIS, the latter being exposed to the liquid and captured prey in the digestive cavity. The mesenteries are arranged radially around the circumference of the body wall and extend along much of its length, partitioning the digestive cavity longitudinally. Mesenteries help to reinforce the body wall, provide sites of digestion, have muscles used by the polyp to retract into the CORALLITE or the COENENCHYME, and contain the gonads. Extrudable mesenteric filaments rich in NEMATOCYSTS extend from the mesenteries and in SCLERACTINIANS are involved in aggressive behaviours against other corals.

mesoglea Gelatinous middle layer between the outer EPIDERMIS and inner GASTRODERMIS of the body wall of a CNIDARIAN.

microbiome The entire suite of mostly unicellular associates (especially bacteria but also algae, fungi and viruses) – the microbiota – living in or on a larger, multicellular host organism such as a coral.

millepore A heavily calcified fire coral, member of the class HYDROZOA. Linnaeus placed the fire corals in the genus *Millepora* (for the many small pores on the surface of the colony) within the LITHOPHYTA.

mitochondrion (*pl.* **mitochondria**) Cellular organelle where the energy stored in food molecules is liberated in cellular respiration when electrons are removed from the molecules and passed to molecular oxygen, O_2.

molecular clock Method for estimating the amount of time since SPECIES or other TAXA diverged during evolution, based on the average rate of accumulation of genetic mutations.

morphology (morphological character) The observable form and structure of an organism.

mutualism Type of SYMBIOSIS beneficial to both partners, for example, the coral host and the ZOOXANTHELLAE it harbours.

MYA Million years ago.

nematocyst Explosive, capsular organelle (any of a number of organized or specialized structures within a living cell) among two dozen or more categories of CNIDAE involved particularly in the envenomation and capture of prey.

ocean acidification Decrease in the pH (decrease in alkalinity, increase in acidity) of seawater due primarily to the dissolution of increasing amounts of atmospheric carbon dioxide. The concomitant decrease in the availability of carbonate ion (CO_3^{2-}) makes conditions less conducive for CALCAREOUS organisms to deposit their skeletons.

Octocorallia (octocorals) Subclass of ANTHOZOA whose polyps have eight radially arranged, unpaired fleshy MESENTERIES and eight branched, feather-like tentacles. Octocorals include precious corals, organ-pipe corals, bamboo corals, blue corals and sea pens, and an array of 'soft corals', sea fans and sea whips. *Compare* HEXACORALLIA.

oxidative stress Ongoing consequences of the imbalance in living with molecular oxygen, O_2, its various REACTIVE OXYGEN SPECIES (ROS) and other cellular oxidants that damage the DNA, proteins and lipids in cells faster than they can be repaired or replaced. *See also* ANTIOXIDANT.

phenotype (*adj.* **phenotypic**) Observable physical or physiological traits of an organism, collectively, brought about by the interaction of its genotype with its environment.

photon A quantum of light energy (or other electromagnetic radiation) that behaves like a particle. The energy of the quantum is inversely proportional to the wavelength of the radiation (for example, long-wavelength, low-frequency red radiation contains low energy whereas short-wavelength, high-frequency ULTRAVIOLET RADIATION contains high, potentially damaging, energy).

phylogeny (*adj.* **phylogenetic**) The evolutionary history of an organism or lineage, including its relatedness to other lineages.

planula (*pl.* **planulae**) Motile (by means of cilia), usually sexually produced, larva in the CNIDARIA.

pluripotent (*adj.*) Term referring to certain undifferentiated STEM CELLS that can give rise to some, but not all, specialized cell types in an organism.

polyp Attached, sedentary or sessile columnar body form of a single cnidarian individual or module. Its body wall of three layers encloses a central digestive cavity having only one external opening (called the mouth, although it serves for both ingestion of food and egestion of waste), surrounded by tentacles used in prey capture or defence.

polypary Skeletal support of a COLONY of polyps, also known as the CORALLUM in the calcified SCLERACTINIANS.

reactive oxygen species (ROS) Chemically reactive molecules containing oxygen alone in different energetic states or degrees of reduction, or in combination with other elements in different degrees of oxidation and reduction. ROS may occur naturally as intermediates or by-products in photosynthesis and respiration, and may be produced by pollutants and radiation, including the ULTRAVIOLET wavelengths in sunlight. The biocidal effects of ROS operate through oxidative damage to DNA and RNA, enzymes and other proteins, and lipids in cellular membranes. Corals have a suite of ANTIOXIDANT and sunscreen defences against OXIDATIVE STRESS.

Scleractinia (scleractinians) Order of HEXACORALLIA comprising hard or stony corals, sometimes called reef-building (or reef) corals.

sclerites Microscopic calcareous skeletal elements produced by OCTOCORALS. Embedded in the MESOGLEA to different extents throughout the COENENCHYME, sclerites add stiffness and support.

Scyphozoa (scyphozoans) Major class in the clade (subphylum) MEDUSOZOA in which a relatively large medusa ('jellyfish') is the sexual stage of the life history. The class diverged early from class ANTHOZOA and contains no members referred to as 'corals'.

septa (*sing.* **septum**) CALCAREOUS blade- or plate-like partitions extending from the wall of the scleractinian CORALLITE towards its centre, radially dividing the corallite. Septa occur in multiples of six, reflecting their deposition by the paired MESENTERIES that originate in cycles of six in each POLYP. This hexamerous symmetry of their septa explains the placement of SCLERACTINIA in HEXACORALLIA.

somatic tissues Body tissues other than those that produce gametes (eggs and sperms) in the gonads (ovaries or testes). Genetic mutations in the DNA of somatic cells may be passed to polyps produced by asexual reproduction, but are not inherited in sexual reproduction, which involves cells of the germ line that become gametes.

species (*sing., pl.*) The most exclusive TAXON, commonly defined as one whose members may freely interbreed with each other but are unable to interbreed and form viable hybrids with members of other species.

stem cell An undifferentiated cell that can perpetuate itself through repeated division and also differentiate into cells specialized for diverse functions, as during embryological development, and during tissue repair, rejuvenation and regeneration of body parts in the adult. In certain hydrozoans, interstitial cells (i-cells) act as stem cells to produce most SOMATIC cell types, such as EPIDERM and GASTRODERM, nerve, muscle, CNIDOCYTES and others; in some, specialized i-cells act as germ-line stem cells that are destined to produce only gametes (eggs and sperms). Interstitial cells have not been described in scleractinians or other anthozoans but evidence points to their presence.

symbiosis The close 'living together' of two or more organisms, often to their mutual benefit. *See also* ENDOSYMBIOSIS; ZOOXANTHELLAE.

systematics Scientific discipline of classifying organisms and elucidating their PHYLOGENETIC relationships.

taxon (*pl.* **taxa**) A unit of organismal classification at any level of a hierarchy. Linnaean hierarchical classification proceeds from the most exclusive taxon, the SPECIES, through the progressively more encompassing taxa – genus, family, order, class, phylum and kingdom – to the most inclusive, the domain. In Linnaean classification each organism has a scientific name known as a binomial, the first part of which is the genus and the second part the species or specific epithet, for example, *Corallium rubrum* or *Acropora palmata* (note the use of an initial capital only in the genus, and italics for both genus and species). The specific epithet is unique for each species in the genus and distinguishes it from the others. Thus, *Corallium rubrum* is the precious red coral found in the Mediterranean Sea, whereas *Corallium konojoi* is a precious white coral from the western Pacific. A more distantly related species of Pacific red coral, *Paracorallium japonicum*, is in a different genus. **Taxonomy** is the sub-discipline of classifying and naming organisms.

telomerase Enzyme that lengthens the TELOMERE, which consists of repetitive DNA sequences on the end of a chromosome.

telomere Terminal cap of repetitive DNA sequences that protects genetic information in the chromosome from being degraded during successive replications of chromosomal DNA.

thorium–uranium (^{230}Th/^{234}U) dating Radiometric technique used to determine the age of calcium carbonate-containing materials such as scleractinian skeletons. Uranium (including the ^{234}U isotope), unlike thorium, is soluble in seawater and can be incorporated into coral skeletons. When the colony dies, no more ^{234}U is added to the skeleton. Over time, ^{234}U decays at a known rate to ^{230}Th (which is initially absent from the skeleton), and the changing ratio of the two isotopes is used to calculate the time when the colony died. The method is useful for materials aged from only a few years up to about 500,000 years old.

ultraviolet (UV) radiation (UVR) At the surface of the earth, a band of solar radiation having wavelengths between 280 and 340 nanometres (nm). UV-B wavelengths shorter than 310 nm are more energetic and potentially damaging than longer UV-A wavelengths. Marine organisms, including symbiotic SCLERACTINIANS, synthesize natural sunscreens that intercept UV-B and UV-A radiation and dissipate the energy harmlessly as heat, before UVR can damage tissues directly or cause the formation of ROS there.

unicellular algae Various single-celled, largely planktonic, photosynthetic organisms (including DINOFLAGELLATES in the genus *Symbiodinium* that also form symbioses with many corals).

vegetative proliferation *See* ASEXUAL REPRODUCTION.

Zoanthidea (**zoanthids**) Order of HEXACORALLIA including the gold corals, among the oldest living animals (some greater than 2,500 years old) and prized for making jewellery.

zoophyte (*arch.*) Zoophyta was one of Linnaeus's two major groups containing corals and included various octocorals and sea pens (and unwittingly a black coral and a scleractinian), more distantly related hydrozoans, and even coralline algae and BRYOZOANS. Although they grew vegetatively and resembled flowers, colonial zoophytes were recognized as composite animals. Others applied the term even more widely to ambiguous TAXA whose members seemed more like plants than animals. *See also* LITHOPHYTE.

zooxanthella (*pl.* **zooxanthellae**) Generic term that refers to several CLADES of DINOFLAGELLATES at the species level (especially those in the genus *Symbiodinium*), which in their asexual, vegetative phase form SYMBIOSES with corals and other marine invertebrates.

zooxanthellate coral Coral (most commonly, a SCLERACTINIAN) that typically harbours symbiotic DINOFLAGELLATES called ZOOXANTHELLAE. Distinguished from AZOOXANTHELLATE CORALS, which do not normally form such symbioses.

Acknowledgements

I envisaged *Where Corals Lie* in 2004 when developing a marine biology course for first-year undergraduates, incorporating illustrative materials from the wider arts and humanities. This book, via those antecedent lectures, benefited from a critical audience of not-yet specialists whom I thank for their interest. The School of Marine Sciences at the University of Maine supported my complementary passions to marine science in my teaching, as I hinted in my 2008 article in *Art Journal*. Its publication, shepherded by Senior Editor Joe Hannan, provided impetus for a book, as did Angela Kingston's subsequent asking of me to write for the catalogue of her curated exhibition 'Underwater'.

My wife Jean Shick and her sister Anne Wood were my editors of first resort and models of the thoughtful non-specialist reader to whom the book especially is addressed. They critiqued early drafts of every chapter and choices of illustrations. Jonathan Burt, editor of the Reaktion Books series 'Animal' for which this volume was originally contracted, helped to prune what was to be simply *Coral* when it ramified too prolifically.

Vicki B. Pearse's wide experiences improved the entire manuscript and helped to realize my vision of this book as something that would engage coral specialists and lay readers alike. Angela Kingston's curatorial eye helped harmonize the many possible illustrations with the text, which she graciously responded to in its entirety. Denis Allemand, Jean-Pierre Gattuso and Peter W. Glynn collectively provided informed critical comments regarding several chapters. All were founts of information beyond my text. Any errors that remain are my own.

At Reaktion Books I thank publisher Michael Leaman for his encouragement, patience and for allowing me to write a larger book than I originally proposed; editor Jess Chandler for her close attention to detail and understanding of my changes to the proofs; and designer Carroll Associates and picture editor Rebecca Ratnayake for supporting my vision of the book and helping greatly in sourcing problematic illustrations at the eleventh hour.

Felipe Paredes translated long passages of Spanish from Monturiol's treatise on his *Ictíneo* submarine. Michael Grillo aided my initial inquiry into the semiotics of red coral and translated from and to Italian for me. Sayoko Mori provided translations and insights regarding Japanese culture, history and religion. Nozomu Iwasaki translated further from the Japanese edition of *Corals: Cultural and Historical Perspectives on Precious Corals* (2011) and helped me obtain some of the images therein to use in my own book. Unless otherwise stated, translations from French and Latin are my own.

Tiffany Filocamo alerted me to the sculpture by Carlo Parlati II of the visage of Medusa in the Museo del Corallo, Ravello, where Tiffany and Giorgio Filocamo welcomed my friend Patricia Dowse, who photographed the sculpture to use in this book. The Filocamos kindly permitted my reusing images from their collection that appeared in 'Parures de la mer' (2000) at the Musée Océanographique de Monaco.

Maurizio Candotti Russo of Idelson Gnocchi Publishers, Ltd graciously allowed me to reproduce images from their publication of the late Capt. Leonardo Fusco's book *Red Gold* (2011).

Many curatorial and library professionals contributed their expertise: James M. Cornelius (Lincoln Collection, Abraham Lincoln Presidential Library and Museum); Ian Fowler (Osher Map Library, University of Southern Maine); Enric Garcia Domingo (Museu Marítim de Barcelona); Pascale Joannot (Muséum National d'Histoire Naturelle); Stefanie Jovanovic Kruspel (Naturhistorisches Museum Wien); Constance Krebs (Association Atelier André Breton); Yuri Long (Rare Books Room, National Gallery of Art, Washington, DC); Laura Martini (Archeologia, Belle Arti e Paesaggio di Siena Grosseto e Arezzo); Alessandra Minetti (Museo Civico Archeologico, Sarteano); Emily Nazarian (Rubin Museum of Art); Diane Rielinger (MBL WHOI Library, Woods Hole); Margit Seebacher and Martina Jurstak (Schatzkammer und Museum des Deutschen Ordens, Vienna); Amy Staples (Smithsonian Institution National Museum of African Art); and Bernard Verlingue (Musée de la Faïence, Quimper). Gregory Curtis (Regional Federal Depository and Interlibrary Loan Department, University of Maine) was especially helpful in locating innumerable, sometimes obscure references, images and recordings.

To help me appreciate the unrecorded music in the 'Coral Riffs' section of Chapter Four, pianist Patricia Stowell performed it for me in her music room. The Music Reference Team at the British Library and Anne Keats of Wirripang Pty, Wollongong, furnished the sheet music that enabled this recital.

ACKNOWLEDGEMENTS

Charlie Veron graciously allowed me to use illustrations from his books and helped to source other photographs. Eric Matson (AIMS) provided his incisive photographs, and with Janice Lough (AIMS), information to put them in context. To illustrate specific points in this book, my valued colleague Éric Tambutté (Centre Scientifique de Monaco) took photographs of corals grown in the aquarium there, patiently allowing me to watch and edit in real time. My good friend and colleague Jean-Pierre Gattuso photographed coral blocks in walls of the old town of Jeddah specifically for this book. Architectural historian Deborah Thompson helped to put the motifs from Korshid's house in Suakin into the larger context of Islamic decorative art. Geologist Mitchell Colgan took the photographs, and he and Peter W. Glynn provided insights regarding the ornamental blocks of fossilized coral used architecturally in Coral Gables, Florida. Patrick V. Kirch kindly provided his photograph of the Taputapuatea temple complex, and the image of the ahu on the Marae Mahaiatea in Tahiti from his personal collection. Vincent Callebaut offered a choice of his visionary architectural images. Anatoly Sagalevich (P. P. Shirshov Institute of Oceanology, Moscow) graciously provided the photograph of RMS *Titanic* made from the MIR-1 submersible. Richard Milner supplied the photograph of Ernest Griset's atoll triptych and Lyulph Lubbock kindly allowed me to use it. Other friends and colleagues also provided their photographs: José Alejandro Alvarez; Andy Davies, Julien Debrueil, Katharina Fabricius, Thomas Guilderson, Ove Hoegh-Guldberg, Peter Huber, Veerle Huvenne, Eunjae Im, Nozomu Iwasaki, Christopher Kelley, Boris Kester, Allison Lewis, Shinichiro Ogi, Mike Qualman, Baruch Rinkevich, Anya Salih, Robert Steneck and Bette Willis.

My friends Vicki and John Pearse helped to negotiate the labyrinth of cnidarian (especially octocorallian) phylogenies, my crude sketch of which Ryan Cowan transformed into what Darwin would recognize as his 'coral of life'. Anne Sheppard kindly let me use her map of the distribution and abundance of coral reefs. Emmanuel Boss and David Townsend helped me create the original maps for this book.

Distinguished contemporary artists, photographers and filmmakers, their associates, and gallery owners generously provided images and permission to use their works for this book. For this and more I am deeply indebted to Daniel Arnoul; Jaq Chartier; Mark Dion and Georg Kargl Fine Arts; Whitney Ganz of William A. Karges Fine Art; Chris Garofalo; Toshiyuki Hara of Hara Shobo Gallery; Lucas Leyva and Coral Morphologic; Courtney Mattison; Alexis Rockman; Zane Saunders; Keiko Schmeisser and Merryn Gates; David Stacey; Roger Steene; Philip Taaffe and Raymond Foye; and Maarten de Wolf.

This book has benefited from my stays at marine laboratories and aboard their ships, and at museums and libraries, immersed in their collections. Several past directors of the Australian Institute of Marine Science (AIMS), Townsville, have supported my many research visits there. The masters and crews of AIMS vessels provided the seagoing support that allowed memorable sojourns among living corals along the Great Barrier Reef.

The leadership of the Centre Scientifique de Monaco (CSM) has supported my coral research there, hosted first by Jean-Pierre Gattuso. My subsequent stays at the CSM (with the support of its president, Patrick Rampal) have been hosted by Denis Allemand (CSM Scientific Director) and Christine Ferrier-Pagès. My work in the architecturally magnificent Musée Océanographique has been a delight fostered by its previous directors François Doumenge and Jean Jaubert, then by Director Robert Calcagno and always by the staff. Anne Marie Damiano and Elisabeth Balsinger helped with my bibliographic requests, and curator Valérie Pisani and photographer Michel Dagnino provided images for this book. Jacqueline Carpine-Lancre, an historian of oceanography responsible for research in the archives at the Palais Princier, took a warm interest in my project and directed me to documents and artworks that otherwise I would not have discovered. In Paris, at the Institut Océanographique, Fondation Albert Ier, Prince de Monaco, Catherine de la Bigne provided access to the collections, and a tour of the art and architecture, for me to photograph.

Many others provided a myriad of contributions too numerous to specify but which, even if I did not use them directly, influenced my thinking and writing here. In particular I thank Nikki Adams, Éric Beraud, Fei Chai, John Dearborn, Richard Defenbaugh, Paul Delaney, James Dykens, Charles Eldredge, Gary Farren, Paola Furla, Philippe Ganot, Diane Genthner, Muriel Gout, Christiane Groeben, Esty Grossman, Jean-Georges Harmelin, Ursula Harter, Stefan Helmreich, Melody Jue, Ronald Kozlowski, Irwin Lavenberg, Todd Leibowitz, Michael Lesser, Sara Lindsay, Yukio Lippit, Michael Mallinson, Tom Martin, Tina Molodtsova, Randy Olson, Stephen Pezzetti, Jacke Phillips, Luisa Piccinno, Paul Rawson, Stéphanie Reynaud, Carly Rustebakke, Robbi Segal, Laura Shick, Lawrence Silver, Susan Spiggle, Gerry Stecca, Shannon Struble, Laurence Talairach-Vielmas, Robert Trench, Wes Tunnell, Paul Tyler, Seth Tyler, Nancy Um, Rhian Waller, Les Watling, Ray Williams, William Zamer, Craig Zievis and Didier Zoccola.

The following institutions generously supported my research and writing about corals: the University of Maine and its School of Marine Sciences; the U.S. National Science Foundation; the Australian Institute of Marine Science; the National Geographic Society; the Centre Scientifique de Monaco; the Musée Océanographique de Monaco; the Bermuda Institute of Ocean Sciences; and the Université de Nice–Sophia Antipolis.

To all I am truly grateful.

Photo Acknowledgements

The author and publishers wish to express their thanks to the below sources of illustrative material and/or permission to reproduce it. Some locations of artworks are given below, in the interests of brevity:

Courtesy Abraham Lincoln Presidential Library & Museum, Springfield, Illinois: 156; © José Alejandro Álvarcz: 213; © Daniel Arnoul: 74, 115, 130, 158; Art Research Library – National Gallery, Washington, DC: 68; Australian Institute of Marine Sciences (AIMS): 37; photo Jean-Gilles Berizzi/Musée du Louvre/© RMN-Grand Palais/Art Resource, NY: 71, 72; from W. G. Bowman, Harold W. Richardson, Nathan A. Bowers et al., *Bulldozers Come First*, 1st edn (New York, 1994): 194; BPK Bildagentur/ Alte Pinakothek, Bayerische Staatsgemaeldesammlungen, Munich/Art Resource, NY: 77; André Breton © 2018 Artists Rights Society (ARD, New York/ADAGP, Paris: 113; © British Library, London: 48, 50, 138, 160; © Trustees of the British Museum: 140, 154; © Vincent Callebaut Architectures: 178; © XL Catlin Seaview Survey: 214, 216; courtesy Jaq Chartier/Elizabeth Leach Gallery: 120; Chester Dale Collection, National Gallery of Art, Washington, DC: 173; © Christie's Images/Bridgeman Images: 103; © Mitchell Colgan: 200, 201; © 2018 Salvador Dalí, Fundació Gala-Salvador Dalí, Artists Rights Society (photo Getty images/ Sherman Oaks Antique Mall/Contributor): 203; from James D. Dana, *Corals and Coral Islands* (New York, 1872): 4; from Charles Darwin, *The Structure and Distribution of Coral Reefs* (London, 1842): 56; © Andy Davies: 100; © John D. Dawson/YSPS: 226; © Julien Debrueil: 109; © Mark Dion/photo Georg Kargl Fine Arts, Vienna: 102; © David Doubilet/National Geographic Creative: 57, 172; from Henri Lacaze-Duthiers, *Histoire naturelle du corail* (Paris, 1864): 9, 53; from Henri Milne-Edwards, *Histoire Naturelle des Coralliaires ou Polypes Proprement Dits*, (Paris, 1857): 10, 19; from John Ellis, *The Natural History of Many Curious and Uncommon Zoophytes* (London, 1786): 3, 176; from *Endeavour* (journal): 34, 118; from Sir Arthur Evans, *The Palace of Minos at Knossos, Volume II, Part II* (London, 1928): 141; Katharina Fabricius/AIMS: 218, 219, 220; © The Field Museum: 82; from Leonardo Fusco, *Red Gold* (with permission from Idelson Gnocchi Publishers, Ltd. (2011)): 166, 169; Galleria Borghese, Rome: 76; © Chris Garofalo: 108; from Andrew Garran, *Picturesque Atlas of Australasia*, vol I (1886): 91; © Jean-Pierre Gattuso: 186, 187; © Rupert Gerritsen: 192; Google Earth/Image © 2017 DigitialGlobe: 197, 198, 199; from Philip Henry Gosse, *A History of the British Sea-anemones and Corals* (London, 1860): 18; from Philip Henry Gosse, *A Naturalist's Rambles on the Devonshire Coast* (London, 1853): 12; from Jurien de la Gravière, *L'Amiral Baudin* (Paris, 1888): 191; Greenwich Museums: 60; © Ernest Grisnet/photo courtesy Richard Milner: 114; © Ove Hoegh-Guldberg: 61, 66; from Ernst Haeckel, *Arabische Korallen* (Berlin, 1875): 42, 62, 123, 184; from Ernst Haeckel, *Art Forms in Nature* (Berlin, 1904): 112; by kind permission of Hara Shobo Gallery, Tokyo: 80; Hawaii Historical Society: 193; from *Histoire de l'Académie royale des sciences avec les mémoires de mathématique et physique* (Paris, 1700): 2; collection of Simone & Peter Huber/photo Peter Huber: 49; © Ralph Hutchings/Visuals Unlimited, Inc: 73; © Eunjae Im: 229, 230; Instituto Nacional de Arqueología e Historia, Mexico: 190; © Nozomu Iwasaki: 79; from George Johnston, *A History of the British Zoophytes* (London, 1838): 11; from *Journal of Marine Research*, vol. 14 (1955)/G. E. Hutchinson: 90; courtesy William A. Karges Fine Art: 174; from *The Coral Reef*, music © Horace Keats/ text © John Wheeler/image courtesy Wirripang Pty Ltd, Australia; collection of Kennedy Museum of Art at Ohio University: 157; from William Saville-Kent, *The Great Barrier Reef of Australia* (London, 1893): 43, 51, 63, 96; © Boris Kester: 181; Wolfgang Kiessling: 217; courtesy Patrick V. Kirch: 179, 180; Kobe City Museum/DNPartcom: 45; © Korea Aerospace Research Institute/European Space Agency, 2013: 189; Kunsthistorisches Museum, Vienna: 131; from *The Coral Reef Are Dreaming Again* (2014), director: Lucas Levya/photo Daniel Fernandez: 209; Allison Lewis: 41, 221; Library of Congress, Washington, DC 133; from Michael Maier, *Atalanta Fugiens* (Oppenheim, 1617): 47; © Justin Mcmanus/The AGE/Fairfax Media/Getty Images: 211; MAREANO/Institute of Marine Research, Norway: 33; courtesy Marine Biological Laboratory | Woods Hole Oceanographic Institution: 95; from *Marine Fisheries Review*, vol. LV (1993)/photo Richard W. Grigg: 14, 171; Eric Mathon/Palais Princier: 231; © 2018 Succession H. Matisse/Artists Rights Society (ARS), New York: 116, 117; © Eric Matson/AIMS: 1, 92; © Courtney Mattison: 207;

courtesy Alessandra Minetti, Museo Civico Archeologico, Sarteano/Laura Martini, Archeology, Fine Arts, and Landscape of Siena Grosseto and Arezzo: 89; Eric Matson, AIMS/details courtesy Janice Lough, AIMS: 26, 27; © Mathurin Méheut 2018 Artists Rights Society (ARS), New York/ADAGP, Paris: 142; Musée du Louvre: 78; courtesy Musée Océanographique de Monaco: 44, 64 and 121 (photos Michel Dagnino), 124; Musée d'Orsay, Paris: 5, 135; Museo del Corallo, Ravello: 69 (photo Patricia Dowse), 86, 87, 88, 132 (photos Michel Dagnino); Museo dell'Opificio delle Pietre Dure, Florence: 128; Museum of Art, Rhode Island School of Design, Providence: 153; © Museum Marítim de Barcelona: 161, 162, 163, 164, 165; Museu Boijmans Van Beuningen, Rotterdam/photo Studio Buitenhof, The Hague: 159; © The Trustees of the Natural History Museum, London: 17; National Museum of African Art, Smithsonian Institution, Washington, DC: 152; Natural History Museum, London/composite illustration originally published in J. Malcolm Shick, *Art Journal*, 2008: 35; courtesy National Museum of Japanese History: 111; Naturhistorisches Museum, Vienna: 97; © NERC, National Oceanography Centre, Southampton/courtesy Veerle Huvenne: 222, 223, 224; © Paul Nicklen/National Geographic Creative: 110; NOAA-Hawaii Undersea Research Laboratory Archives: 15; from *Oceanography*, vol. XX (2007): 225; © Shinichiro Ogi: 167, 170; courtesy Osher Map Library, University of Southern Maine: 93, 94; Palazzo Vecchio, Florence: 75; Parent Géry (Wikimedia Commons): 6; © 2018 Estate of Pablo Picasso/Artists Rights Society (ARS), New York: 134; © Luisa Piccinno: 146; from Erik Pontoppidan, *Natural History of Norway* (Lyon, 1755): 59; private collection/photo © Christie's Images/Bridgeman Images: 85; photo Mike Qualman/courtesy Integrated Orbital Implants, Inc: 208; from Eugène de Ransonnet, *Sketches of the Inhabitants, Animal Life and Vegetation in the Lowlands and High Mountains of Ceylon, as Well as of the Submarine Scenery near the Coast, Taken in a Diving Bell* (Vienna, 1867): 122; © RMN-Grand Palais/Art Resource, NY: 127, 129, 137; © Alexis Rockman: 206; from Aristide Roger, *Voyage sous les flots* (Paris, 1869): 177; from Leonard Rosenthal, *The Kingdom of the Pearl* (New York, 1920)/image courtesy Bromer Booksellers, Inc: 106; from Louis Roule, *Description des Antipathaires et Cérianthaires recueillis par S.A.S. le Prince de Monaco dans l'Atlantique nord. (1886–1902)*, (Monaco, 1905): 81; Rubin Museum of Art, New York: 46, 143, 145; Anatoly Sagalevich, Head of deep manned submersibles of P. P. Shirshov Institute of Oceanology, Moscow: 36; Anya Salih/Confocal Facility, Western Sydney University: 65 (pair); courtesy SARS Greenlaw Archive, GRE E410: 188; © Zane Saunders: 205; Schatzkammer und Museum des Deutschen Ordens, Vienna: 147, 150; drawing SD-5511 Linda Schele, © David Schele: 55; © courtesy of Estate of Jörg Schmeisser: 31, 233; from Alexander von Schouppé, *Courier Forschungsinstitut Senckenberg*, vol. 164 (1993): 83; photo Shai Shafir/courtesy Baruch Rinkevich: 232; from *Shakespeare's Comedy of the Tempest with Illustrations by Edmund Dulac* (London, 1908): 234; Malcolm Shick: 24, 29, 52, 70, 104, 212, 215; © David H. Stacey: 204; Peter Stackpole/The LIFE Picture Collection/Getty Images: 99; © Robert Steneck: 28; © Philip Taaffe: 32; Eric Tambutté/Centre Scientifique de Monaco: 22, 23, 39, 40, 175; © Jason deCaires Taylor. All rights reserved, DACS/ARS/Artimage 2018: 105, 227; from David W. Townsend, *Oceanography and Marine Biology* (2012)/permission of Oxford University Press, USA: 58; University of Iowa Libraries, John Martin Rare Book Room: 125; University of Queensland Library: 1821; US Coast Guard: 195; from Jules Verne, *Vingt milles lieues sous les mers* (Paris, 1871): 98; © J.E.N. Veron: 38; from J.E.N. Veron, *Corals of the World*, vol. 1, 2000/artwork Geoff Kelley: 16, 25; 67, 119 (photos Roger Steene); © Victoria and Albert Museum, London: 148, 155; Wellcome Library, London: 54; © Pete Souza/White House images: 84; © Bette Willis: 20, 21; © Maarten de Wolf: 144

Bin im Garten, the copyright holder of image 151, LoKiLeCh, the copyright holder of image 126, and Shahee Ilyas, the copyright holder of image 210, have published them online under conditions imposed by a Creative Commons Attribution-Share Alike 3.0 Generic License; Martin Falbisoner, the copyright holder of image 183, and J. Miers, the copyright holder of image 202, have published them online under conditions imposed by a Creative Commons Attribution-Share Alike 4.0 Generic License

Readers are free:

> to share – to copy, distribute and transmit the work
> to remix – to adapt this image alone

Under the following conditions:

> attribution – You must attribute the work in the manner specified by the author or licensor (but not in any way that suggests that they endorse you or your use of the work).

> share alike – If you alter, transform, or build upon this work, you may distribute the resulting work only under the same or similar license to this one.

Index

Illustration numbers are indicated by *italics*.

Abbott, Tony, Prime Minister of Australia 237–8
Acanthaster planci (crown-of-thorns sea star) 29, 230–31, *212*
acclimatization to environmental change 247–9, 256, 285
Acheulian toolmakers 91
Acropora 23, 195, 250, *21*
    A. humilis 125, *112*
    A. hyacinthus (table coral) 248–9
    A. millepora 256
    A. palmata (elkhorn coral) 123, 125, 244, 290, *113*
acroporid corals 105, 229, 230, 231, *212*
Actiniaria (order) 17, 285, 287
    see also sea anemones
adaptation (evolutionary process) 246–7, 248, 249, 256, 285
Africa
    East (Sudan; Tanzania) 159, 160, 198, 201–2, 267, *Map 2*
    North (Algeria; Libya; Morocco; Tunisia) 24, 84, 156, 159, 173, 177, 178, 179, 180, 182, 266, *168*, *Map 1*
    South Africa 236, *215*
    sub-Saharan 165, 178
Agassiz, Alexander 62, 112
Agassiz, Louis 62
ageing 29, 69
Agricola, Georgius, *De natura fossilium* 48
airfields 187, 209, 211, 212, 213, *195*, *196*, *199*
al-Idrisi, Muhammad 160, 173
Al-Tur, Sinai 129, 132–3, 202, 266, 267, *121*, *123*, *184*, *Map 1*, *Map 2*
Albert I, HSH Prince of Monaco 111, 139
Albert II, HSH Prince of Monaco 256–7, *230*
Alborean Sea 182, 266, *Map 1*
Alcyonacea (order; alcyonaceans) 15, 17, 24, 30, 51, 54, 189, 241, 285, 286
Alcyonium digitatum (dead man's fingers) 15, 17, 55, *11*, *12*
Alcyonium palmatum (dead man's hand; robber's hand) 43–5, *36*

Aldred, Cyril, *Jewels of the Pharaohs* 157
Alexandria 157, 178–9
Algeria 178, 180, 266
Allemand, Denis 256–7, *231*
Amalfi 178
American Samoa 248, 253, 256, 269, *Map 4*
Amsterdam, coral trade in 179, 266, *Map 1*
amulets, corals as 89, 92, 93–4, 95, 219, *83*, *85–8*, *89–90*
Andersen, Hans Christian, *The Little Mermaid* 105–7, *100*
angel skin coral *see Corallium secundum*
animalcules 54, 58–9, 66, 116, 187
anoxia 20
Antarctic corals 63
Anthozoa (class; anthozoans) 12, 14–15, 17, 42, 57, 63, 65, 70–71, 227, 285, 287, 288, 289, *8*
Anthropocene period 219, 222, 230, 231, 285
antioxidants 29, 69, 71, 234, 285, 289
Antipatharia (order; antipatharians) 17, 18, 43, 45, 89, 91–2, 136, 139, 157, 183, 285, 287, *14*, *81*, *125*, *128*
    see also black coral
Antipathes erinaceus (a black coral) 90–91, *81*
Antipathes griggi (=*Antipathes dichotoma*; a black coral) 18, 183, *14*, *171*
antipathin 18, 285
Aqaba, Jordan 156, 266, *Map 1*
aquarium 24, 33, 45, 63, 111, 117, 145–6, 149, 169, 182, 192, 215, 252, 256–7, 261, *44*, *231*
Arabian Peninsula 202, 203, 267, *Map 2*
Arabian/Persian Gulf 89, 157, 159, 236, 246, 253, 267, *215*, *Map 2*
aragonite 17, 20, 30, 48, 51, 123, *49*
    saturation state in seawater 239, 241–2, 244, 285, *221*
architects and builders, corals seen as 98–9, 104, 187–9
Arcimboldo, Giuseppe, *Water* 142–3, *131*
arhats 153, 157, *143*

Aristotle, *History of Animals* and *The Parts of Animals* 43
Arnold, Augusta Foote, *The Sea Beach at Ebb Tide* 44–5, *42*
Arnoul, Daniel
    Chevalier 139, 141, 143, *130*
    Doigt de Mars (Finger of Mars) 80, 83, *74*
    Self-portrait 125–6, 127, *115*
artificial and enlarged islands
    Fiery Cross (Yongshu) Reef 212–13, *268*, *197*, *198*, *199*, *Map 3*
    Maldives 213
    San Blas archipelago 194–5
    Spratly Islands 212
    Tern Island 209–11, *195*
asexual reproduction and vegetative proliferation 12–14, 24, 29, 36, 56, 250, 285, 286, 287, 289, 290, *7*
Astrolabe 97, 100, 101, *95*
atholhu 228
atmospheric CO_2 32, 227, 238–41, 258, 262
    see also carbon dioxide (CO_2); fossil fuels
atolls 10, 59, 62, 72, 95, 98, 104, 112, 115, 126–7, 130, 150, 188, 208, 209, 212, 225, 226–9, 285, 288, *58*, *114*
Attenborough, Sir David 263
azooxanthellate coral 35, 231, 242, 286, 290
bacteria 29, 42, 231, 233, 244–6, *225*
Bahamas 123, 269, *Map 4*
Bali, Hindu demon king 7, 86–7
Bali 253, 256, 268, *229*, *230*, *Map 3*
Ballantyne, Robert Michael, *The Coral Island* 128, 149, 188–9
bamboo corals 17, 18, 37, 178, 225, 288, *13*
banking (and trade in corals) 179
Banks, Sir Joseph 50–53, 57–8, 96–7, 105, 172, 194–5, 261, *48*, *50*, *160*, *179*
Barbados 207, 269, *Map 4*
Barbarikon, Pakistan 157, 267, *Map 3*
Barbary Coast 178
Barcelona 173, 178, 179, 266, *Map 1*
barra italiana 173, 182
barrier reef 59, 62, 97, 241, 288, *56*, *58*
Barygaza, India 157, 267, *Map 2*
Bastion of France 178

Batavia (Dutch East Indiaman) 95–6, 208–9, *192*
Bates, Marston and Donald P. Abbott, *Coral Island* 126
Bavaria 92, 160
beads of coral 46, 92, 94, 149, 153–4, 157, 158, 159, 160, 165–6, 169, 171, 180, *136*, *144*, *151*, *152*, *153*
HMS *Beagle* 14, 59, 104, 126, 188, 219, *114*
Beijing 159, 268, *Map 3*
Belize 58, 196, 236, 269, *55*, *215*, *Map 4*
Benin, kingdom of 84, 165–6, 168, *152*
Benin bronzes 168, *151*
Berenike, Egypt 201, 266 *Map 1*
Bermuda 207, 236, 269, *215*, *Map 4*
bioerosion 31–4, 62, 214, 228, 230, 240–41, *27*, *29*, *30*, *31*
Bio-Eye® orbital implant 225–6, *208*
bioherm 24, 286
Biorock™ 195, 252–6, *299*, *230*
bivalve molluscs 31, 33 *27*, *29*
black band disease 245
black coral 17, 18, 29, 43, 45, 49, 69, 109, 136, 139, 150, 156–7, 159, 178, 183, 285, 287, 290, *14*, *128*, *143*, *171*
  see also Antipatharia; antipatharians
Blake, William 98, 116–17, *107*
bleaching *see* coral bleaching and mass coral bleaching
blood 7, 43, 48, 77, 79, 80–87, 115, 126, 166, 217, *70*, *71*, *73*, *74*
blue corals (order Helioporacea) 17, 288
Boccone, Paolo 47, 54
Bockscar 212
bodhisattvas 157–8
Bonaparte, Joseph, king of Naples and Sicily 179
bone grafts 225
boom-and-bust industry 153, 182
botanical and horticultural image of corals 7, 43, 48, 51, 63, 115, 116, 117, 188, 120–22, 189, 261, *59*, *108*, *110*
HMS *Bounty* 144
Bougainville, Louis-Antoine de, French explorer 126
Bourdon, Sébastien, *The Liberation of Andromeda* 86, *77*
Boyle, Robert 123
Bradbury, Roger, 'A World Without Coral Reefs' 219
brain (meandrine) corals 7, 25, 26, 124, 125, 196, *24*, *112*
Brandt, Karl 65
Breton, André, 115, 123, 125, 215, *113*

Brisbane 207, 268, *Map 3*
broadcast spawning 23, 41 *20*
brooding (of larvae) 24, 41, 248, 250
Bryozoa (phylum) 63, 286, 290
Buddhism 88, 153, 157–9, 162, *145*
budding of polyps 12–13, 14, 24, 25, 26, 27, 29, 56, 250, 285, *7*
Buffon, Georges-Louis Leclerc, Comte de *Histoire naturelle, generale et particuliere, avec la description du Cabinet du Roy* 55
Burgess Shale fossils 15
calcification 11, 12, 14–15, 29, 32, 79, 105, 107, 139, 263, 285, 286, 288–9
  enhancement in corals by endosymbiotic algae 41, 68
  and pH; aragonite saturation 68, 239–40, 241, 242–3
  see also calcium carbonate; ocean acidification
calcite 15, 20, 23, 285, 286
calcium (Ca^{2+}) 15, 175, 239, 286
calcium carbonate (CaCO$_3$)
  accumulation and accretion in coral reefs, balance between erosion and dissolution 31–2, 33, 188, 228, 243
  crystalline 11, 15, 17, 27, 68, 123, 239, 285, 286, *25*
  saturation state in seawater 239
Calicut, India 159
Callebaut, Vincent 193–4, *178*
calyx 15
Cambrian period 14, 15, 240, *217*
Cane, Arabia 157, 267, *Map 2*
Cape Creus 173, 266, *161*, *Map 1*
  see also Monturiol, Narcís
Cape Tribulation 96, 98–9, *91*, *93*, *94*
Cape York 99, 236, 238, *94*
carbon dioxide (CO$_2$)
  and carbon cycle 239
  emissions from burning of fossil fuels 227, 239, 241–2, 257, 262
  fixation in photosynthesis 66
  greenhouse gas and global warming 227
  and ocean acidification 32, 239
  and ocean pH 240, 241–3, *218*, *219*, *220*
  reef community respiration 241
carbonate chemistry 239–40
carbonic acid 66, 239
Caribbean Sea 97, 112–3, 123, 125, 250–53, 269, *113*, *105*, *227*, *Map 4*
coral disease 243–4, 245

coral reef decline 219
coral rock as building material 187, 198, 200–201, 207–8, 213, *183*
poaching of reef fishes in 231, 232–3, *213*
rising sea level in 228
sea fans and sea whips 79
shifting baselines in condition of coral reefs 230
threshold temperature for coral bleaching 236, *215*
carotenoids 70
Catala, René 69–70
  Carnival under the Sea 70, 150
Catalans 173, 175, 217
Çatalhöyük 154
Catalonia 173, 178
Catedral de Santa María la Menor, Santo Domingo 198, 200–201, *183*
Cathedral of Our Lady of Peace, Honolulu 198
Cellini, Benvenuto, *Perseus with the Head of Medusa* 80–81, 84, *70*
Celts 154, *140*
Cerianthara (order; tube anemones) 14, 17, 286, 287, *8*
Cervantes, Miguel de, *La Galatea* 145
Ceuta 160, 178, 179, 266, *Map 1*
Ceylon (Sri Lanka) 70–71, 101–3, 267, *64*, *97*, *121*, *Map 2*
Chagos Archipelago 212, 267, *Map 2*
Chartier, Jaq, *Great Barrier Reef* 132–3, *120*
chimera 7, 40, 53, 54, 118, 250, 286 *108*
China
  artificial islands 187, 212–13, *197*, *198*, *199*
  coral fishing 153, 178
  coral poachers 184
  Han dynasty 47
  precious red coral reaches 47, 157
  protection of precious corals 183–4
  Qing dynasty, red coral featured in court hierarchy 159–60
  red coral symbolism 88, 145
  Song dynasty 159
  Tang dynasty 159
  Western Han dynasty 47, 157
chitin 18
chlorophyll 41, 63, 65, 66, 71, *40*
Christmas Island 238, 269, *Map 4*
chromoproteins (CPs) 70, 71, 287
CITES (Convention on International Trade in Endangered Species) 181, 183, 184

clade 41, 235–6, 249–50, 286, 288–290
Clarke, Arthur C., *Dolphin Island* 70, 97
climate change (global)
    corals and reefs 10, 30, 184, 219, 238
    coral bleaching 41, 72, 234, 257
    public awareness of 222–5, 262, 285, *206, 207*
    responses of reef corals to 231, 246–50, 253, 257, 262
    sea level rise 226–7, 228–9, *211*
'climate change physiology' 249
clones 24, 26, 250, 286, 288
cnidae 11, 286
    see also nematocysts
Cnidaria 11–12, 14, 29, 63, 117, 285–9, *8, 59*
coenenchyme 12, 15, 45, 286, 288–9, *9*
coenosarc 25, 27, 286
cold-water corals 21, 63, 231, 243, *18, 59*
Cole, Michael 80
Coleridge, Samuel Taylor 56–7, 188
collagen 15, 18
colonialism 7, 126, 222
colonies
    advantages of 30, 35–6, 182, 229, 249–50
    form or morphology and size of 7, 14, 19, 20, 29, 30, 37, 39, 40, 48, 51, 56, 124–5, 150, 181, 285, *16, 26, 38, 112*
    formation and skeletal support of 12, 15, 18, 19–20, 24, 25–6, 27, 30, 31, 43, 56, 286, 287, 289, *7, 9, 12, 25*
    longevity of 18, 26, 30, 290
colours of reef corals 66–7, 69–72, 73, *63, 65, 66*
colours and optics of coral reef waters 126–7, 128–34, *114, 117, 118, 119, 120, 123*
Columbus, Christopher 198, 207
commensalism and mutualism 68
Compagnie Marseillaise du Corail 179
Compagnie Royale d'Afrique 179
concretions 7, 47, 48, 51, 189, 192
Cook, James 7, 57–9, 91, 96, 97, 98–100, 105, 115–16, 187, 194–5, 219, 222, 241, 261, *93, 94*
Cooktown, Australia 101, 112, *96*
COP21 (21st Conference of the Parties to the United Nations Framework Convention on Climate Change) 229, 236, 262
corailleurs 180
coral and bells 143–4, *132, 133*

coral bank 20, 59, 72, 77, 126, 178, 180, 182, 192, 212
coral bleaching, mass coral bleaching 7, 10, 41, 71, 72, 73, 109, 230, 233–7, *1, 66, 214, 215, 216*
    description and first accounts of 233–4
    high-temperature anomalies and El Niño 227, 233, 234, 237
    repetitive 234, 247, 257, 258
    sunlight and high temperature synergy 41, 233, 234
    threshold temperature 234, 236, *215*
coral calcium 225
Coral Castle, Florida 214, 216–17, *202*
Coral Gables, Florida 213–15, *200, 201*
coral gardening (reef restoration) 252
Coral Merchants Street, London 179
Coral Morphologic 150–51, 227
coral nurseries (reef restoration) 258, *232*
coral of life, the 11, 14, *8*
coral reefs
    built on the ruins of their past 193, 207
    as castles 187, 189, 190–92, *175, 176*
    as classical architectural and Atlantean ruins 192–3
    commodification and monetization of 263
    coral (areal) cover on 219, 247
    dangers of 95–8, 105
    drugs from the sea 222
    ecosystem services provided by 242, 263
    global distribution and biodiversity of 19, 20, *17*
    as graveyards 100–105, *95, 97, 98, 99*
    'pristine' 230
    resilience of 233, 245, 249, 253, 257
    restoration of 248, 252–6, 262, 263, *228, 229, 230, 232*
    sustainable urbanity 193
    zombie ecosystems 219, 256
Coral Sea 32, 104, 187, 189, 192, 268, *Map 3*
Coral Triangle 19, 20, 212, 286, *17*
Coralliidae (family of precious corals) 15, 286
Corallimorpharia (order; corallimorpharians) 14, 17, 20, 242, 286, 287, *8*
corallite 24, 25, 27, 30, 39, 55, 91, 93, 100, 125, 203, 205, 286, 288, 289, *23, 25, 112, 188*
    see also starrystone

Corallium (genus of precious corals)
    C. elatius 168, 177
    C. konojoi 178, 290
    C. rubrum 11, 12–13, 15, 42–3, 45, 48, 118–19, 154, 156, 176–7, 181–2, 290, *6, 7, 9, 109*
    C. secundum 177, 178, 183
    see also red coral
cores (drilled) 26, 28–9, 30, 62, *26*
cornuto 80, 94, *87*
Coromandel Coast, India 144, 158–9, 267, *Map 2*
Coronado, Francisco Vázquez de 169
Corsica (Mediterranean Sea) 178, 266, *Map 1*
Cortés, Hernán 207–8
Cretaceous period 23, 240, *217*
Crete 156, 266, *Map 1*
Cro-Magnon Aurignacian culture 92
'Crochet Coral Reef' 225
crushed coral 187, 209, 211–13, *194, 195, 196, 198, 199*
crustose coralline algae 31, 32, 35, 230, *28*
Cryogenian period 11
cryptic species 40, 286
cryptochromes 24
crystals, corals envisioned as 48, 51, 112, 115, 117, 123–5, 150, 217, 261, *48, 112*
curiosity cabinets 47, 51, 86, 115, 134, 136–7, 139, 143, *125, 126, 128*
cutlery 86, 162, *148*
Cuvier, Georges 11, 47, 105
Daikoku 180–81, *170*
Daintree Rainforest 222
Dalí, Salvador 173, 215, 216–7, *203*
Dana, James Dwight, 8, 100, *4*
Darwin Point 59, 62
Darwin, Charles
    and Agassiz, Alexander 62
    Beagle voyage and diary 14, 59, 126–7, 219, *114*
    'coral architects' 104, 188
    'the coral of life' 14, *8*
    and Lyell, Charles 59, 104
    natural selection 246
    Structure and Distribution of Coral Reefs, The 59, 62
    studies at Cambridge 122
    subsidence theory of atoll formation 59, 62, 100, 104, 188, *58*
Darwin, Erasmus, *The Botanic Garden* 116
Dash, Mike, *Batavia's Graveyard* 96, 208–9, *192*

INDEX

deep-sea corals and mounds 35, 36, 63, 68, 111, 219, *33, 59, 104*
Dendrophyllia ramea 105, 134, *100*
desert, tropical ocean as 10, 68–9, 72
Diadema antillarum (sea urchin) 231
diagenesis 31, 287
diamonds 179
Dickens, Charles 189
Dicquemare, abbé Jacques-François 57
Diego Garcia naval base 212, 213, 267, Map 2
digestion 12, 29, 33, 36, 42, 66, 230, 288, 289, *7*
dinoflagellates 40–1, 65, 68, 235, 246, 285, 286, 287, 290, *39, 40, 41*
Dion, Mark, *Bone Coral (The Phantom Museum)* 109, *102*
Oceanomania project 136, 139
Dioscorides, Pedanius 11, 43, 47, *De materia medica* 43
diseases of corals 41, 227, 233, 243–4
defences against 245
microbial pathogens 244–6, *225*
white pox in *Acropora palmata* 244, *225*
Disney, Walt, *20,000 Leagues Under the Sea* 105, *99*
diving suits 112, 129, 134, 173, 176, *161, 162*
DNA 29, 69, 152, 249, 289, 290
Dobbs, David, *Reef Madness: Charles Darwin, Alexander Agassiz, and the Meaning of Coral* 59, 62
Doctorow, Cory, 'I, Row-Boat' 105, 261
Dominican Republic 198, 233
Donati, Vitaliano 55–6
Donnelly, J. J., 'Sea-Wraith' 98
Doumenge, François 178–9
Drake, Francis 198, 207
dredging 20, 111, 207–9, 211–13, 237, 261, *104, 198*
Dry Tortugas, Florida 112
Ducie, baron Francis 144, *133*
Dulac, Edmund
Birth of the Pearl 114–15, *106*
Full fathom five thy father lies 108–9, 110, *101*
And deeper than did ever plummet sound I'll drown my book 264–5, *234*
Dumont d'Urville, Jules 7, 97, 100–101, 111, *95*
Eakin, Mark 236
East India Company 159, 179
Easter Island 236, *215*

Eastern Pacific Ocean 109, 233, 269, 287, Map 4
Ebisuya, Konojo, father of Japanese precious coral industry 178
echinoderms, 18
crinoids (sea lilies) 8, 9, *5*
crown-of-thorns sea star (*Acanthaster planci*) 29, 230–31, *212*
brittlestars 31, 35, *32*
Eco, Umberto, *The Island of the Day Before* 77, 105, 123, 145, 189
Ediacaran period 11, 15
Edo people 165
Edo (Tokugawa) period 89, 168–9
Edom (Idumea) 156
Egypt 143, 156–7, 171, 201
Ehrenberg, Christian Gottfried 57, 123
Eilat, Israel 156, 258–9, 266, *232*, Map 1
El Niño 109, 229, 233, 237, 256, 287
El Tor *see* Al-Tur
elevated volcano hypothesis 59, 188
Elgar, Edward, 'Where Corals Lie' 112, *149–151*
elkhorn coral (*Acropora palmata*) 123, 125, 244, 290, *113*
Ellis, John 8, 53, 54, 56, 57
The Natural History of Many Curious and Uncommon Zoophytes 51, 56, 57, 192, *3, 176*
Elworthy, Frederick Thomas, *The Evil Eye: An Account of This Ancient and Widespread Superstition* 93
Emmert, Paul, *View of the Honolulu Fort—Interior* 209, *193*
Emperor seamounts 178, 268, Map 3
end-Permian mass extinction 15, 20, 23, 240
HMS *Endeavour* 57–61, 91, 96, 99, *57, 91*
Endeavour Reef 96, 98, 222, *93*
endosymbiosis 20, 26–7, 37, 41, 65, 66, 68–9, 71, 233, 246, 249–50, 287, 289, *25, 39, 40*
Enewetak Atoll, Marshall Islands 62, 236, 268, *215*, Map 3
Environmental Protection Agency (EPA), U.S. 238, 257
engin 52–3, 172, *50, 160*
Enola Gay 211–2, *196*
environmental stress 233–4, 245–7, 248–50, 286
epidermis (ectoderm) 11–13, 27, 286–9, *7, 25*
epigenetics 249, 287
Erdoğan, Recep Tayyip 202
Erebus, Mount 63–5, *60*
Eternal Reefs, Inc. 219, 252–3, *228*

etymology of 'coral' 11
eugenics 250, 287
evil eye 93, 94, 158
evolution and phylogeny of corals 11, 14–15, 17–20, 22, 24, 28, 40, 219, *8*
exoskeleton 12, 18, 287
Explanaria (= *Turbinaria*) 118, 120–21, 122, *110*
Ezekiel (Book of) 156
fascinum 93–4, 143, *85*
Ferdinand, archduke of Tyrol 162, *149*
Ferdinand I (of the Two Sicilies) 179
Ferdinand II, king of Aragón 179
Fiery Cross (Yongshu) Reef 212–13, 268, *197, 198, 199*, Map 3
filter feeding by corals 36–7, 39
fire corals (*Millepora*) 14, 287, 288, *8*
fish 10, 31, 32, 33, 35, 36, 57, 58, 122–3, 128, 195, 196, 222, 231, 232–3, 241, 242, 247, 252, 256, 258–9, 263, *33, 55, 213, 226, 230, 232*
fishing, effects on corals and reefs
commercial 231
destructive (by explosives and poison) 33, 35, 261, 262–3
illegal 35, 261, 262
overfishing 195, 230, 231, 261
reduced fishing pressure 256
spear-fishing 176, 231, 232–3, *213*
subsistence 263
sustainable 233
trawling, damage by 20, 168, 178, 243, 244–5, *223, 224*
Flaubert, Gustave
fleurs de fer 7, 48, 123
The Temptation of Saint Anthony 7, 48, 57, 105, 123
fleurs du corail 47, 51, 54, 105, 117, *53*
Flinders, Matthew, 29, 98, 105, 115–16, 122
circumnavigation of 'Terra Australis' and charting of 'Great Barrier Reefs' 59
description of formation of a coral cay 72
naming of Great Barrier Reef 59, 115
notion of coral 'architecture' 187–8
Voyage to Terra Australis, A 72
floral image of corals 14, 57, 58, 63, 65, 100, 109, 115, 116, 117, 123, *55, 59, 62*
Florida Keys 214
flos ferri 48, 51, 123, *48*
Flourens, Marie-Jean-Pierre, French physiologist 55, 56
flower animals 14, 57, 123, 285
see also Anthozoa

fluorescence 287
  by chlorophyll 41, *40*
  by corals 41, 69–70, 71, 132, 150, *40, 64, 65*
  by fluorescent proteins 41, 69, 71, 132, *40, 64*
  see also green fluorescent protein
fluorescent proteins *see* fluorescence; antioxidants
HMS *Fly* 26, 53, 101, 118, 146, 189
Forsskål, Peter 201
fossil fuels 227, 239, 241, 262
Four Tantras 158
France 111, 263
  marine station at Arcachon 146
  and Mediterranean red coral 48, 156, 178, 180, 182
  Pacific explorations 97
Francis, Pope, 'On Care for Our Common Home' 262
free-diving 97, 171, 173, 176
French Frigate Shoals Airfield, Tern Island 209–11, *195*
Freneau, Philip, 'The Hurricane' 98
fringing reef 59, 62, *58*
Fungia ('mushroom' coral) 53, *51, 52*
Fusco, Leonardo 176–7, *Red Gold: Extreme Diving and the Plunder of Red Coral in the Mediterranean* 176
Gaimard, Joseph Paul 58, 59, 97, 100
Galaxea fascicularis 189, 190–92, *175, 176*
Galle I, Cornelius, *Fishing for Coral* 172, *159*
gametes (egg and sperm) 12, 23, 29, 41, 285, 289
Gans, Johann Ludwig, *Coralliorum historia* 48
Gardiner, J. Stanley 66, 68
Garnett, Richard, 'Where Corals Lie' 7, 112, 149, 151
Garofalo, Chris, *Corallum Desertus* 118, *108*
gastrodermis 11–12, 27, 41, 66, 286–9, *7, 25*
Geddes, Patrick 65–6
gene expression 248–9, 287
genetic mosaics, coral colonies as 250
genetic rescue of corals 248
Genoese 169, 178, 179
geographic (latitudinal) range shifts of corals 247
geographically separated populations 246, 248
Gilgamesh, king of Uruk 89, 134
Glynn, Peter W. 234

gold corals (order Zoanthidea; zoanthideans) 18, 19, 183, 184–5, 287, 290, *15, 172*
  Hawaiian species *Kulamanamana haumeaae* 19, 183, 184–5, *15, 172*
Golijov, Osvaldo, 'Coral del Arrecife' 152
gonads 12, 23, 287–9, *7*
Gordon, general Charles George 201
Goreau, Thomas J., 252
Gorgon 7, 77, 80, 146, 171, 261, *158*
gorgoneion 79, 86
gorgonians
  abyssal 36–7, 38–9, *36*
  Antarctic 63
  artistic representations 110–11, 139–40, 152, 154–6, 189, *103, 104, 142*
  as lithophytes 51
  as plants (former classification) 49, 51
  precious corals and other octocorals, relationship to 15, 45, *14*
  sea fans, sea whips 15, 79, 245, *10, 17, 127*
  as Zoophyta (zoophytes) 56
gorgonin 15, 17–18, 49
Gosse, Philip Henry 63, 128, 132, 188–9
  Actinologia Britannica: A History of the British Sea-anemones and Corals 21, 117, *18*
  A Naturalist's Rambles on the Devonshire Coast 17, *12*
  paintings by 21, 129–30, *18*
Great Barrier Reef (GBR)
  in the arts 37, 98, 123, 130, 132, 133, 193, 218–19, 219–22, 225, 260–61, *34, 118, 120, 204, 205, 233*
  Australia's national passion for 237, 261
  British Association for the Advancement of Science's Expedition 37, 66, 104, 129, *35*
  coral cover on, decline 219
  crown-of-thorns seastar 230–31, *212*
  HMS *Endeavour* 57–8, 98–9, *93*
  freshwater flood events 31, *27*
  as 'The Labyrinth' 96, 99, 100, 115, 219, *94*
  management of 219
  mass bleaching 6–7, 233–4, 236, 247, 256–7, *1*
  mass spawning of corals 23, *20, 21*
  naming by Flinders 59, 115
  ocean acidification and reef community calcification 241

  threats to 233, 237–8, 247, 261
  UNESCO World Heritage Area (WHA) 237–8
Great Barrier Reef Marine Park Authority (GBRMPA) 257
green fluorescent protein (GFP) 41, 69, 71, 132, 287, *40*
greenhouse gases 227, 233, 236, 239
Grigg, Richard W. 183
Griset, Ernest, untitled watercolor triptych of an atoll 126–7, *114*
growth of corals
  branching (staghorn) and foliose corals 229–30, 241, 244, *221*
  heliotropic 39, *37*
  indeterminate 24, 39, 287
  light availability, effects on 37, 39, 248
  plasticity of 39–40, *38*
  record of, in skeleton 30, 214–15, *201*
  reef corals, growth-forms of 19, *104, 16*
  reduced growth in bleached corals 72, 233
  symbiotic algae and bacteria 37, 39, 42, 66
  vegetative (budding) 12–14, 24, 29, 36, 56, 250, 285, 287, *7*
  and water flow 36–7, 39
Guam 211, 236–7, 249, 268, *216, 226, Map 3*
Guangzhou 159, 268, *Map 3*
Gulf of Mexico 187, 207, 269, *Map 4*
Ha'amonga 'a Maui (Maui's Burden) trilithon 196, 198–9, *182*
Habsburg imperial family 136, 162, *149*
Haeckel, Ernst
  Arabische Korallen 44–5, 63, 65, 133–4, 202, *42, 62, 123, 184*
  continuity of life, and corals as crystals 123
  Kunstformen der Natur 123–5, *112*
  light and dark, coral's ability to distinguish 63
  Tur (Al-Tur), Sinai 129, 134, 202, *123, 184*
Hall, James W., *Bones of Coral* 109
Hallstatt Iron Age sites 154
Hamilton-Paterson, James 7, 56, 97, 193, 262–3
Hamwass, Zahi, *King Tutankhamun: The Treasures of the Tomb* 157
hard corals *see* reef corals; Scleractinia
Hawaiian mythology 91, 196
Hawaiian Islands 198, 208, 236, 256, *215*

Hawaiian precious and semi-precious corals
  black (*Antipathes griggi*) 18, 183, *14*, *171*
  gold (*Kulamanamana haumeaae*) 18, 19, 184–5, *15*, *172*
  pink (*Corallium secundum*) 183
heat stress *see* environmental stress
Heredia, José-Maria de, 'Le Récif de corail' 117
hermaphroditic corals 23, 287
Hexacorallia (subclass; hexacorals) 15, 17, 18, 25, 124–5, 139, 285, 286, 287, 288, 289, 290, *22*, *23*, *112*
  see also orders of Hexacorallia: Actiniaria; Antipatharia; Ceriantharia; Corallimorpharia; Scleractinia; Zoanthidea
Hilbertz, Wolf 252
Hiroshima, atomic bombing of 212, 261
Hishikawa, Harunobu, *A Scene of the Folk Tale Momotarō, Depicting Momotarō's Return with Treasures after Slaying Daemon* 89, *80*
hive mind 188
Hollande, François, president of France 256–7, *231*
Holmes, Richard, 104, 219, 263
holobiont 41, 287
Honolulu
  churches of coral rock 198
  fortification of harbour by Kamehameha I 208, 209, *193*
Hopkins, Gerard Manley, 'Binsey Poplars' 230
Hormuz 46, 159, 267, Map 2
hotspots
  coral biodiversity 222
  coral bleaching 234
  coral disease 243
Houtman, Frederik de, Dutch explorer 95
Houtman Abrolhos, Western Australia 95, 208–9, 268, *192*, Map 3
Hughes, Robert
  Barcelona 175
  The Fatal Shore 208
human-assisted evolution of corals 250, 264
human-assisted migration of corals 248
Hutchinson, G. Evelyn 80, 84, 95, *90*
Huxley, Thomas Henry 55, 65
hydrogen ion (H^+) 68
hydroxyapatite (in human bone) 225

Hydrozoa (hydrozoan corals) 14, 287
iace (beads of black coral) 45–6
Ictíneo, Ictíneo II 174–6, *163*, *164*
immortality (extreme longevity) 18, 26, 27, 29, 88, 98, 107, 143, 183, 225
Imperato, Ferrante 49, 51, 134
India, precious coral trade 46, 153, 154, 157, 159, 171, 178, 179
Indian Ocean 267, 268, Map 2, Map 3
  Arnhem Land 89, 268, Map 3
  damage to ships by corals 43, 95
  Diego Garcia (U.S. naval base) 212, 267, Map 2
  Houtman Abrolhos 95, 268, Map 3
  Maldives 227–8, 236, *215*
  mass coral bleaching in 234
  quarrying of coral for construction 205
  reef coral diversity 20, 253, *17*
  Seychelles 231, 233, 267, Map 2
  Silk Road, maritime 157, 201
  South Africa 236, *215*
  Stone Town, Zanzibar 198, 267, Map 2
Indonesia 268, Map 3
  Ambon island 49, 122, 136
  Bali and Biorock™ restoration projects 253, 254–5, 256, *229*, *230*
  human reliance on coral reefs 219
ingegno 172, 173, 176, 178, 180, *160*, *169*
Institut Océanographique de Paris 111
intelligent design 250, 288
Intergovernmental Panel on Climate Change (IPCC) 219
International Energy Agency 257
International Society for Reef Studies 262
HMS *Investigator* 115
Isabella I, queen of Castile 179
Islamic culture 160, 178, 201–5
Italy 94–5, 169, 177, 168, 180–81
Iwasaki, Nozomu 184
Jamnitzer, Wenzel, *Daphne* 139–40, 143, *129*
Jeddah, Saudi Arabia 202, 204, 206–7, *186*, *187*, *189*
jewellery 18, 183, 285, 290
  see also red coral
Jews 178–9
Johnson, Samuel, *A Dictionary of the English Language* 90
Johnston, George, *A History of the British Zoophytes* 8, 10, 17, *11*
Jones, Steve, *Coral: A Pessimist in Paradise* 69, 193, 207

Jukes, Joseph Beete 26, 29, 30–31, 53, 118, 101, 122–3, 128, 146, 189, 196
  Narrative of the Surveying Voyage of HMS Fly, . . . in the Torres Strait, New Guinea, and Other Islands of the Eastern Archipelago, in the Years 1842–1846 146
Jussieu, Bernard de, French naturalist 54
Kachur, Lewis, *Displaying the Marvelous: Marcel Duchamp, Salvador Dalí, and Surrealist Exhibition Installations* 215
Kagoshima Prefecture 177, 182, 268, Map 3
Kamehameha I, king of Hawaii 197–8, 208–9
Kawaguti, Siro 69
Kawaiaha'o Church (Honolulu) 198
Keats, Horace, and John Wheeler, 'The Coral Reef' 150–51, 193, *139*
Keratoisis flexibilis (a bamboo coral) 18, *13*
Key Largo limestone (Keystone) 214–15, *200*, *201*
Key West (Cayo Hueso, 'Bone Key') 109
Khorshid Effendi house 202–3, 205, *185*, *188*
Kirch, Patrick, *A Shark Going Inland Is My Chief* 196
Kiribati 118, 228–9, 237, *110*, *211*
Kitchener, Horatio Herbert 201
Klunzinger, Carl Benjamin, *Upper Egypt: Its People and Products* 112, 123
Kōchi Prefecture (Japan) 88, 89, 177, 182, 268, *79*, Map 3
Kōin, Nagayama, *Inro, Turtle Netsuke, and Coral Bead* 166, *153*
kouralion 11, 43, 79
Kroeber, Alfred L. 188
Krukenberg, C.F.W. 65
Kulamanamana haumaeaae (Hawaiian gold coral) 19, 183–5, *15*, *172*
Kuna people 194–5, 228
La Calle, Algeria 178–9, 267, Map 1
La Tène culture 154, *140*
Lacaze-Duthiers, Henri, *Histoire naturelle du corail* 15, 54, 153, 180, *9*, *53*
Laccadives, Indian Ocean 68, 267, Map 2
lace corals (Hydrozoa) 14, 287, *8*
Laforgue, Jules, 'se madréporiser' 117–18
lagoons 10, 62, 104, 118, 126–30, 146, 188, 202–3, 209, 228, 285, 288, *58*, *114*, *116*, *185*

Lamarck, Jean-Baptiste 11, 15, 189
Lamu, Tanzania 201
Land of Punt 201
Lapérouse, Comte de (Jean-François de Galaup) 97, 100–101, 104, *95*
Laval, François Pyrard de 48, 95, 105
Lavoisier, Antoine, French chemist 65
Leduc, Alphonse, 'Le Collier de corail' 149
Lévy-Dhurmer, Lucien, *Méduse, ou Vague furieuse* 146, 148–9, *137*
Leyva, Lucas, *The Coral Reef Are Dreaming Again* 226–7, *209*
Lincoln, Mary 168–9, *156*
Ling, Xu, 'Wanzhuan-ge' 145
Linnaeus, Carolus (Carl) 51, 56–7, 201
  Systema naturae 47
  taxonomy of Lithophyta and Zoophyta 8, 56, 288, 290
Lisbon, coral trading in 179
Lithodendron ('stone tree') 43, 49, 53
Lithophyta 8, 56, 288
lithophyte ('stony plant') 7–8, 51, 53, 55, 57–8, 105, 111, 136, 138–9, 189, 288, *2, 104, 127*
Liverino, Basilio, *Red Coral: Jewel of the Sea* 80
Livorno 179–80, 266, *Map 1*
Lizard Island 130, 236, 241, 268, *118, 215, Map 3*
London 169, 179, 192, 266, *Map 1*
London, Jack 134
Longfellow, Henry Wadsworth, 'To a Child' 143–4
Lophelia (*pertusa*) 21, 35–6, 46, 63, 111, 124–5, 244–5, *18, 33, 59, 104, 112, 222, 223, 224*
Low Isles 66, 233
Lowe, Percy, *A Naturalist on Desert Islands* 72
Lowell, Percival 134
Lyell, Charles 59, 104, 112
  Principles of Geology 59
madrepore (*Madrepora*; hard or stony coral) 7–8, 51, 53, 59, 70, 91, 105, 115, 117, 122, 123, 189, 192, 288, *3*
Magdalenian culture 92, *82*
Maghreb, the 159–60, 178, 266, *Map 1*
Maier, Michael 48–9, 149, *47*
  Atalanta fugiens 48, 49, *47*
Maldives 68, 230, 236, 267, *215, Map 2*
  coral used for construction 201, 213
  rising sea level 227–8
Malé 213, 228, *210*
Mangin, Arthur, *The Mysteries of the Ocean* 117

mano fico 94, *86*
Mantegna, Andrea, *Madonna della Vittoria* 80, 82, 83, *71, 72*
marae Mahaiatea (Tahiti) 194, 195, *179*
marae Hauvivi (Ra'iatea) 196, 197, *180*
marine carbon sink 239
marine protected areas (MPAs) 183, 231, 233, 246, 249, 262
Marsa'el Kherez (now El Kala) 160, *178*
Marseille 47, 153, 156, 178, 179, 266, *Map 1*
Marsigli, Luigi Ferdinando 48, 51–3, 153, 182–3, *50*
  bathymetry of red coral distribution 171
  fleurs de corail 51, 54
  Histoire physique de la mer 51–3, 153, 171–2, *48, 50, 160*
Martin, Marguerite Marie, *The Flowery Rock and Pink Wrasses* 44, 45, *44*
Martin, Paul Bartholomew 179
mass extinctions 20, 240, *217*
mass spawning of reef corals and its synchronization 23–4
materia medica, corals as 43
Matisse, Henri 128–30, 132, 134, 149
  The Apataki Channel 128, *116*
  lagoon at Fakarava 128, 132, 134, 146, *116*
  Le Lagon 128
  Polynesia, The Sea 128–9, *117*
  Tahitian lagoon sojourn 10
Matthiessen, Peter, *Far Tortuga* 97
Mattison, Courtney 222, 224–5
  Our Changing Seas II 224, *207*
Maui 18, 183, 196, *171*
Maui (Polynesian hero) 91, 196
Maya 57, 58, 196, *55*
Mayor, Alfred Goldsborough 112
medicinal use of corals 43, 48, 89, 93, 225
  see also *materia medica*
Mediterranean Sea 11–12, 15, 43, 45, 47, 80, 83–4, 88, 110, 139, 153–4, 156–7, 159–60, 165, 168–9, 171–3, 176, 178–82, 242, 266, 290, *6, 45, 74, 160, Map 1*
medusa (jellyfish) 12, 15, 53, 54, 129, 285, 288, 289, *117*
Medusa (mythological Gorgon) 7, 76–81, 84, 86–7, 107, 146, 148–9, *68, 69, 70, 75, 77, 78, 79, 137*
Medusozoa 14, 15, 287, 288, 289, *8*
Méhèut, Mathurin, *Service, La Mer* 154–56, *142*

Méliès, Georges, *Deux Cent Milles [Lieues] Sous les Mers, ou, le Cauchemar du Pêcheur* 105, 193
Melville, Herman 128
  Clarel 77, 98, 109
  Mardi 107
  Omoo 105, 117, 118, 146, 189
  Timoleon, Etc., with 'Fruit of Travel Long Ago', including 'Venice' 187, 193
Merian the Elder, Emblem XXXII in *Atalanta fugiens* 48, 49, *47*
mesenteries 12–13, 24, 36, 287, 288, 289, *7*
mesoglea 11, 12, 13, 27, 286, 288, 289, *7, 25*
Miami 150, 213–15, 226–7, 269, *200, 201, Map 4*
Michelet, Jules, *La Mer*, 63, 65, 97, 118, 146, 189, 192–3
microbiome of corals 42, 249, 288
Middle Ages 143, 178
Midway corals 183
Midway Island 178, 211, 269, *Map 4*
Mignard, Pierre, *Perseus Liberating Andromeda* 86, 87, *78*
millepore (hydrozoan coral) 7, 8, 288, *4*
Millepora (hydrozoan 'fire coral') 14, 288
Milne-Edwards, Alphonse 111
Milne-Edwards, Henri, *Histoire naturelle des coralliaires ou polypes proprement dits* 16, 17, 22–3, 47, *10, 19*
Milsom Jr, Charles, *The Coral Waltzes* 149, *138*
Mino, Jacopo di, *Madonna col Bambino* 95, *89*
Minoan culture 154, 156, *141*
Mitchell, Joni 212–13
Mitchell, Stuart, 'Coral Fugue' 152
mitochondria 29, 288
molecular clock 11, 14, 15, 288
Momotarō the Peach Boy (Japanese folk tale) 88, 89, *80*
Monet, Claude, *Palazzo da Mula, Venice* 186, 187, *173*
Montgomery, James, *The Pelican Island* 26, 98, 99, 213
  criticism by James Dwight Dana 100
Monturiol, Narcís 173–6
Mo'orea 195
Moreau, Gustave, *La Galatée* 8–9, *5*
Morocco 182, 266, *Map 1*

INDEX

Moulton, Septima, 'The Coral Mazurka' 149
Muscatine, Leonard 68
Musée Océanographique de Monaco 45, 139
Muséum National d'Histoire Naturelle, Paris 189
music 146, 149–52
mutation, genetic 11, 29, 249, 250, 288, 289
mycosporine-like amino acids (MAAs) 68, 69, 71
Nagasaki and Hiroshima, atomic bombing of 212, 261
Naples 169, 173, 178–80, 266, *Map 1*
Nasheed, Mohamed, president of The Maldives 228
Natternzungenkredenz (*languier*; *Natterzungenbaum*; tonguestone holder) 160–62, *147*
Nature Conservancy, The 233
Neanderthal Mousterian site 91
nematocysts 11, 14, 27, 286–8, *25*
see also cnidae
Nereids 8, 84, 107, *8*
Neuville, Alphonse de, 'Everyone Knelt in an Attitude of Prayer' (in Jules Verne *Vingt milles lieue sous les mers*) 104, *98*
New Caledonia 69, 234–5, 268, *214*, *Map 3*
Niebuhr, Carsten 187, 201
Nishidomari Tenmangu shrine 88–9, 176–7, 180–81, *79*, *167*, *170*
nitrogen-fixing bacteria 42
NOAA (National Oceanic and Atmospheric Administration, U.S.) 236, 258
Northeast Canyons and Seamounts National Monument 231
Nusa Kunda (Skull Island, Solomon Islands) 197–8, 268, *181*, *Map 3*
nutrients in seawater 26, 30, 36, 68, 72
Oahu 183, 184, 230
Oba (king) of Benin 84, 165–6, 168, *152*
Obama, Barack, former U.S. president 93, 230, 231, 238, 257, *84*
O'Brian, Patrick, *The Thirteen-gun Salute* 97
ocean acidification 10, 20, 32, 230, 238–9, 240–43
Octocorallia (subclass; octocorals) 12, 14–15, 17, 23, 29, 30–31, 43, 44–5, 54, 139, 178, 183, 285–286, 287, 288, *7*, *8*, *42*, *43*
odour of corals 53–4, 196

Odum, Eugene P., *Fundamentals of Ecology* 72
Okinawa 182, 225, 268, *Map 3*
Olalquiaga, Celeste, *The Artificial Kingdom: A Treasury of the Kitsch Experience* 86, 192–3
One Tree Island 241
oolite 31, 213, 214–15, 227
Opuhala 91
organ-pipe coral (*Tubipora musica*) 43, 44–5, 51, 136, 149, 157, 288, *42*, *43*, *125*
Oro war cult 196, 197, *180*
Ovid (Publius Ovidius Naso), *Metamorphoses* 54, 77, 79, 80, 107
oxygen, molecular (O_2)
　life-giving gas 65
　menace 68–9
　respiration 29, 245, 288, 289
　production in photosynthesis 65, 66, 68, 289
　supply and depletion of 20, 36, 66, 68, 175, 176, 239, 245
　see also reactive oxygen species
oyster reefs, likened to coral reefs 263
Ozama Fortress, Santo Domingo 207
Pacific Remote Islands Marine National Monument 209
Pacific species of precious corals
　beginnings of the industry 177–8
　boom-and-bust cycles of industry 153, 182
　and CITES 182, 184
　(selective) collecting by submersibles 182, *183*
　colors of various species 177
　export to Italy 181
　as a family business 181
　fishing gear 176, 177, 178, *167*
　management, preservation and protection 182–3
　poaching of 184
　scuba depth (below) 178, 183
　see also precious corals; red coral
　value of 181, 184
Palau (Micronesia) 69, 236, 268, *215*, *Map 3*
　natural acidification 241–2, 244, *221*
Palmyra Atoll 209
Palumbi, Stephen 253, 256
Pandora Reef, Great Barrier Reef 7, *1*
Papua New Guinea 236, 268, *215*, *Map 3*
　in Coral Triangle 20, 286, *17*
　mass coral bleaching 236, *215*

volcanic seeps and naturally acidified reef sites 241–3, *218*, *219*, *220*
Paracelsus, *Book of Vexations* 47–8
Paracorallium japonicum 177, 182, 184, 290
Paris Agreement 229, 238–9, 241, 257, 262–3
　see also COP21
Parlati, Carlo 11, *Medusa* 78–9, *69*
parrotfishes 33, 231–3, *29*, *30*, *213*
pearls 46, 48, 84, 107, 114–16, 157, *106*
Pepys, Samuel, 'A Pleasant Jigg Between Jack and His Mistress' 144
Percival, James Gates 117, 145
Periplus of the Erythraean Sea 157–8, 202
Perseus 7, 76–7, 79–81, 84, 86–7, *68*, *70*, *75*, *77*, *78*,
Peter the Great, tsar of Russia 136
Petoskey stone 92–3, *84*
Peyssonnel, Jean-André 53–5
　account by Huxley, T. H. 55–6
　living corals, observation of 54, *53*
　Traité du corail 43
pH 68
　and aragonite saturation state in seawater 241, 242
　low pH effect on corals and calcification 240–41, 242–3, *218*, *219*, *220*, *221*
　sites of naturally low pH caused by high CO_2 240–41
　see also ocean acidification; hydrogen ion (H^+); carbon dioxide (CO_2)
Philippines 212, 262–3
　position in Coral Triangle 19, 268, 286, *Map 3*
Phillips Brothers 168–9, *155*
Philpot, Glyn Warren, *Under the Sea* 110–12, *103*
Phoenicians 156, 266, *Map 1*
photoinhibition 235
photons 37, 289
photooxidative stress 69, 233–4
photosynthesis 20, 39, 65–6, 68, 71, 234, 248, 289
phylogeny 14–15, 17, 29, 289
Picasso, Pablo, *Humorous Composition: Jaume Sabartés and Gita Hall May* 144, *134*
Pisa 143, 178
Pitcairn island group 144
plankton (phyto- and zoo-) 10, 24, 36, 39, 68, 231, 239, 242, 248, 250, 258, 287, 290

301

planula larva 12, 24, 31, 36, 250, 289, *7*
Pliny the Elder (Gaius Plinius Secundus) 11, 43, 45–7, 51, 93, 95, 153–4, 157, 171
poaching
  of corals 182–4
  of coral reef fishes 232–3, *213*
Pocillopora damicornis 40, 250, *38*
Polo, Marco 158–9
Pollitt, Arthur W., *The Coral Island* (six musical sketches for piano) 149
Polynesia (Polynesian) 77, 89, 91, 107, 128–9, 187–8, 196–7, *117*
  French 107, 196, 269, *Map 4*
polyp 7, 11–14, 15, 17, 19, 23–7, 29–30, 35–6, 39, 41, 43, 47, 49, 51, 53–8, 63, 65–6, 70, 77, 80, 91, 98, 100, 104, 107, 117, 118–19, 125, 126, 146, 187, 188, 189, 213, 225, 231, 236, 242, 249–50, 262, 285–9, *7, 9, 12, 22, 25, 39, 40, 109, 212*
polypary (*polypier*) 7, 55, 57, 189, 192, 287, 289
polypus 56, 80, 98
Pontoppidan, Erik, *Natural History of Norway* 46, 63, *59*
Pool, Matthys, engravings in Marsigli's *Histoire Physique de la Mer* 50–53, 172, *48, 50, 160*
Port Sudan 202
Portier, Paul, *Physiologie des animaux marins* 117
Portugal (Portuguese) 95, 165, 179, 266, *Map 1*
poulpe (octopus; polyp) 12, 43, 55
Powys, John Cowper, *Porius* 118
precious corals (octocorals; Corallidae) 15, 23, 286, 288
  Mediterranean 7–8, 11, 24, 43, 48, 77, 79, 88, 107, 115, 136, 139, 146, 149, 153–4, 157, 159, 165, 171, 176, 182, 183, 184, 219, 290, *128, 138, 140, 145, 152, 166*
  Pacific 89, 153, 168, 176–8, 180–84, 290, *167, 170*
Priestley, Joseph, FRS, English chemist 65
'pristine' coral reefs 230
Pritchard, Zarh, 134–5, *124*
  Coral Arches 186–7, *174*
Provençals 173, 266, *Map 1*
Puerto Rico 151, 236, 244, *215*
Queensland 268, *Map 3*
  Cape York 99, 236, 238, *94*
  coal mining 237–8
  'Crochet Coral Reef' 225

government of 237–8
Great Barrier Reef 100, 130, 218–22, 241, *118, 204, 205*
  industrialization of coastline 237
  Queensland Cement and Lime Company (Brisbane) 207
Quinet, Edgar, *La Création* 192
Quoy, Jean-René-Constant 58–9, 97
Rabelais, François, *Gargantua and Pantagruel* 144–5
Radiata 11
Ra'iatea 196–7, *180*
Rainbow Serpent 222
rainforests 10, 222
Ransonnet-Villez, baron Eugen von 69, 104, 132–4
  Ceylon Coastal Undersea Landscape 101, 102–3, *97*
  Coral Reef at Tor near the Port Entrance, 133, *121*
  Sketches of the Inhabitants, Animal Life and Vegetation in the Lowlands and High Mountains of Ceylon, as Well as of the Submarine Scenery near the Coast, Taken in a Diving Bell 133, *122*
  Submarine Rocks with Green Corals 69, 70, 71, *64*
reactive oxygen species (ROS) 27, 69, 71, 234, 285, 289
Réaumur, René-Antoine Ferchault de 54–5
  Mémoires pour servir à l'histoire des insectes 55
recovery (of coral reefs) 184, 245, 247, 250
recycling (of nutrients in symbiotic corals) 72, 193, 194, 222
red coral 12–13, 15, 52–4, 118–19, 153–4, 158–9, 176, *6, 7, 9, 50, 53, 109, 145, 166*
  amulets 93–5, 143, 153, *85, 86, 87, 88*
  conservation and protection 176, 182, 183, 184
  fishing and harvesting 153, 156, 168, 171–3, 176–8, 180, 182, 184, *159, 160, 167*
  jewellery 45, 47, 122, 146–7, 149, 153–4, 157–8, 160, 168–71, 184, *111, 136, 144, 155, 156, 157, 158*
  myths of origin 46–7, 76–81, 84, 86–8, *46, 68*
  poaching of 182, 183, 184
  religious art and artefacts 80, 82–3, 88, 94–5, 153, 158, 160, 169–71, *71, 72, 89, 146*

Red Sea
  antipatharian (black) corals 45, 156–7
  architectural use of coral 187, 201, 202, *185*
  corals and reefs 23, 43, 65, 112, 123, 149, 157, 202, 253, 258–9, *184, 232*
  ports and towns 129, 156, 201–2, 206–7, 266–7, *189, Map 1, Map 2*
  trade and merchants 157
Red Sea Style (architecture) 201–5, *186, 187, 188*
Redon, Odilon
  Fleur de Sang 144–6, *135*
  interest in marine life 145–6
reef balls 252–253, *228*
reef corals (scleractinians)
  association with death and bones 101–5, 107–10, 196–8, 226–7, *97, 98, 99, 101, 102, 209*
  building and construction material 187, 194–215, *179, 180, 181, 182, 183, 185, 186, 187, 188, 190, 192, 193, 194, 195, 196, 197, 198, 199, 200, 201*
  myths of origin 89, 91, 107
  prehistoric use of 91–2, *82*
reef gaps 20, 239–40, 242, 263, *217*
regeneration 29, 107, 229, 289
Remotely Operated Vehicle (ROV) 177
Remps, Domenico, *Still Life in Deception* or *Cabinet of Curiosities* 139, 143, *128*
Renaissance 47, 80, 95, 117, 136, 160
Rengade, Jules (Roger, Aristide), *Voyage sous les flots* 189, 192, *177*
resilience (of coral reef communities) 233, 245, 249, 253, 257
resistance (of individuals to environmental stress) 41, 71, 243, 245–6, 248, 250
respiration
  cellular and organismal 29, 68, 288–9
  community 241
Rinkevich, Baruch 250, 252
Roberts, Callum, *The Ocean of Life* 109
Rockman, Alexis, *The Pelican* 222–3, *206*
Romans 11, 79, 93–5, *111*, 153, 156–7, 159, 171, 193, 201
Ross, James Clark, RN, explorer 63
Rossetti, Dante Gabriel, *Monna Vanna* 146–7, 184, *136*
Roule, Louis, *Antipathaires et Cérianthaires* 90–91, *81*

Rugosa (rugose corals) 20, 22–3, 30, 92–3, 286, *19*, *84*
Rumphius, Georg Eberhard 49, 53
  D'Amboinsche Rariteitkamer 51
  Amboinsche Kruidboek 51
runoff (terrestrial and agricultural) 30, 231, 233, 237–8, 243, 245, 261
Ruysch, Frederik, *Thesaurus Animalium Primus/Het Eerste Cabinet der Dieren* 136, *125*
Ryukyu Islands 182, 241, 268, *Map 3*
St Andrew's Cross 173, 180, 182, *169*
Sahara Desert 245
Sala, George Augustus 169
salabre 52, 53, 173–5, 182, *50*, *163*
San Blas archipelago (Panama) 194, 228, 269, *Map 4*
San Juan de Ulúa fort 207–8, *190*, *191*
Sardinia 24, 176, 178, 266, *Map 1*
Saunders, Zane, *Coral Serpent* 219–21, *205*
Saville-Kent, William
  Manual on the Infusoria 66
  The Great Barrier Reef of Australia 44–5, 53, 65–7, 101, 194, *43*, *51*, *63*, *96*
Schmeisser, Jörg
  Some Fragments on the Beach 34, 35, *31*
  Under the Sea, Great Barrier Reef 260–61, *233*
Schleiden, Matthias Jakob 128
  Das Meer 189
  The Plant: A Biography 123
Sciacca, Sicily 180–81
Scleractinia (order; scleractinians) 14, 19–20, 23–4, 27, 29, 40–42, 65, 68–9, 233–6, 248, 286, 290, *8*, *25*, *39*, *40*
sclerites 15, 17, 289, *12*
scuba diving 10, 97, 105, 176–8, 182–3, 230, *166*, *171*
sea anemones 8, 14, 17, 20–21, 29, 43, 54–5, 57, 63, 65–6, 69, 110, 117, 225, 227, 285–7, *8*, *18*, *209*
sea fans and sea whips 14–17, 30, 79, 116, 138–9, 152, 245–6, 288, *8*, *10*, *127*, *225*
sea levels 62, 91, 153, 193, 195–6, 205, 214, 222, 226–30, 248, *210*, *211*
sea pens (order; Pennatulacea) 14, 15, 288, 290, *8*
sea surface temperature (SST) 227, 234, 247, 287
seawater, pH of 23, 32, 54, 68, 239, 241–4, 247, 288 , *218*, *219*, *220*, *221*

seaweeds (macroalgae) 7, 36, 49, 66, 68, 77, 79, 84, 86, 146, 156, 230–31, 241, 248
septa 22–5, 53, 203, 286, 289, *19*, *23*, *52*
Seychelles 48, 95, 231, 233, 267, *Map 2*
Shakespeare, William 107, 109, 145, 225, 264
shanhu 47, 88, 157, 160
sharks' teeth 160–62, *147*
shifting baseline syndrome 230, 231, 241, 264
shogunate 89, 166, 168–9, 177
Sicilians 47–8, 94, 160, 171–3, 178–9, 180, 266, *Map 1*
Silk Road, maritime 157, 158, 201
Simplicio, Dan
  'Branch Coral and Carved Turquoise Necklace' 170–71, *157*
skeleton (of corals)
  bone grafts and orbital implants, material for 225, 226, *208*
  composition and structure of 11, 15, 17, 18, 20, 23–5, 27, 30–31, 178, 285, *9*, *10*, *12*, *23*, *25*, *27*
  secretion and growth of 12, 15, 24–6, 30, 68, 188, 242
  ^{230}Th/^{234}U dating of 30, 195, 290
  X-ray analysis of 30
  see also growth of corals
skull (human) 102–4, 109, 139, 143, *97*, *128*
Slessor, Kenneth, 'Seven Visions of Captain Cook' 261
Society Islands 195, 196, 269, *Map 4*
soft corals (alcyonaceans) 14, 15, 36, 43–4, 54–5, 132, 193, 241, 285, *11*, *12*, *33*, *44*, *121*
Solander, Daniel 57
solar heat (trapping in atmosphere and oceans) 227
solar power 10, 193
solar radiation 69, 290
somatic (cells, tissues and mutations) 30, 250, 289
South China Sea 187, 212, 268, *Map 3*
Southeast Asia 35, 202, 219
Southey, Robert, 'The Curse of Kehama' 100, 117
spatial competition among scleractian corals 36
spirochaetes (bacteria) 42
Spondylus (thorny oyster) 57, 152, 169
Spratly Islands 212
  see also Fiery Cross (Yongshu) Reef
Stacey, David, *Living in Paradise* 218, 219, 222 *204*

staghorn corals 39, 72, 73, 229, *66*
Star II submersible 183, 184, *104–5*, *172*
starrystone 91–2, 203, 205, *188*
Steinbeck, John, and Edward F. Ricketts, *Sea of Cortez* 32, 35
stem cells 27, 29, 30, 289
Stephenson, Thomas Alan 37, 104, 129, 130, 134, *34*, *35*, *118*
Stevenson, Robert Louis 24, 189, 192
Stewart, Matthew, *Monturiol's Dream* 175
Stone Town, Zanzibar 198
Stott, Rebecca, *The Coral Thief* 23
Strabo, *Geography* 43
stress *see* environmental stress
Stylophora pistillata 25, 29, 41, *22*, *23*, *39*, *40*
Suakin, Sudan 201–3, 205, 267, *185*, *188*
submarines 101, 105, 175, 176, 193, 212
submarine springs 240–41
subsidence theory of atoll formation 59, 62, 100, 104, 188, *58*
Sudan 160, 201–2
sunburn and sunscreens 68, 71, 289, 290
Surrealists and coral 105, 123, 125, 193, 215, 216, 217, *133*, *203*
Swahili Coast (East Africa) 198, 201, 267, *Map 2*
Symbiodinium 41, 42, 235, 290, *41*
  S. thermophilum 246
  see also dinoflagellates; endosymbiotic algae; zooxanthellae
'symbiont shuffling' 236
symbiosis 289
  see also endosymbiosis; zooxanthellae
symmetry 11, 23, 25, 91, 203, 289, *19*, *22*, *23*
Taaffe, Philip, *Sea Stars with Coral II* 35, *32*
Tabarka Island, Tunisia 84, 178, 179, 266, *168*, *Map 1*
Tabulata (tabulate corals) 20, 23, 30, 286
Tahiti 10, 91, 128, 134, 146, 194, 195, 236, 269, *179*, *215*, *Map 4*
Taira no Tomomori 166–9, *154*
Taiwan 153, 178, 181, 183, 184, 268, *Map 3*
Takemitsu, Toru, *Coral Island (An Atoll)* 150
Tanzania 198, 201
Taputapuatea temple complex 196, 197, *180*
taxonomy 11, 15, 40, 49, 288, 290

Taylor, Jason deCaires 111
  Grace Reef 250–51, *227*
  Vicissitudes 112–13, *105*
telomeres 29–30, 290
tentacles 11, 12, 15, 17, 25, 27, 36, 41, 53, 66, 70, 118, 287–9, *7, 22, 25, 39, 40, 51, 109*
Tescione, Giovanni
  Il corallo nelle arti figurative (Coral in Figurative Art) 157
  Il corallo nella storia e nell'arte (Coral in History and Art) 84, 156
thangkas 157
Theophrastus 11, 47
  De lapidibus and *Historia plantarum* 43
Thynne, Anna 24
Tibet 43, 46, 88, 153, 157–9, 161, 179
Tikal 196, 269, *Map 4*
Tiffany & Co. 184
Tinian 211, *196–7, Map 3*
RMS *Titanic* 36–9, *36*
Tokugawa Shogunate 169, 177
tombs 26, 99–100, 104–5, 157, 171, 196, 201, 219, 252, *98*
Tong, Anote, President of Kiribati 228
Tongatapu 196, 199, 269, *182, Map 4*
Tor *see* Al-Tur (also called El Tor and Tur, Sinai)
Torre del Greco 179–81, 266, *Map 1*
Torres Strait (far North Queensland, Australia) 97, 101, 268, *Map 3*
tourism 23, 97, 184, 213, *14*, 219, 222–33, 242, 261, 263
Tournefort, Joseph Pitton de 8, 143, *2*
  Éléments de botanique 49
Trapani, Sicily 160, 178, 266, *Map 1*
trawling 20, 178, 243–5, *223, 224*
Tree of Life (biblical) 94, 115, 126–7, 139, 158, *115*
trees and forests, corals as 7, 17, 43, 45, 47, 48, 51, 56, 88–9, 95, 107, 112, 117, 122, 139–41, 143, 153, 156, 159–60, 182–3, 189, *128, 129, 130, 143*
Treharne, Bryceson, and Zoë Atkins, 'Corals: A Sea Idyll' 149
Trembley, Abraham 29, 54, 56
Triassic period 15, 20, 240, *217*
Trump, Donald J., U.S. President 238, 257, 262, 264
Tunisia 84, 177, 179, *168, Map 1*
Turnbull, Malcolm, Prime Minister of Australia 238
Tutankhamun 157, 171
Tuvalu 228, *Map 3*

ultraviolet (UV) radiation 68–71, 248, 250, 289, 290
Urashima Tarō 88, 89, *79*
U.S. Navy Seabees 209, 211
Utagawa, Kuniyoshi, *Taira no Tomomori* 166–7, *154*
Vallayer-Coster, Anne, *Sea Fans, Lithophytes and Seashells* 138–9, *127*
Vanikoro 97, 99–101, 104, *95, Map 3*
Vanis, Michail 193–4
Vasari, Giorgio, *Perseus Freeing Andromeda* 84, *75*
Venus de Milo 111
Veracruz, Mexico 207–8, *190, 191, Map 4*
Verne, Jules, *Vingt milles lieues sous les mers* 97, 101, 104–5, 122, 146, 149, 176, *98, 99*
Veron, J.E.N. (Charlie)
  atmospheric CO_2 and mass extinctions 239
  Corals of the World 19, 27, 130, *16, 25, 119*
  A Reef in Time: The Great Barrier Reef from Beginning to End 97–8, 262
Victorian era 7, 24, 117, 149, 168–9, 187–9, *155*
Villa-Lobos, Heitor, 'Coral (Canto do Sertão)' 151
viruses 42, 240, 245, 287, 288
Vishnu 86–7
volcanoes and seeps 59, 62–3, 65, 188, 243
Voltaire (François-Marie Arouet), *Dictionnaire philosophique* 56
Wallace, Alfred Russel, *The Malay Archipelago* 122
water quality 238, 243, 256
Wells, Herbert George, *The War of the Worlds* 192
West Wallabi Island, 208–9, *192*
Western Pacific Ocean 229, 256, 268, *Map 3*
Williamson, J. E. 105, 123, 125, *113*
Wilson, W., *Great Morai of Temarre at Pappara in Otaheite* 194–5, *179*
Wolf, Maarten de, *Yemeni Dress 15* 158, *144*
World War, First 110
World War, Second 176, 187, 208–9, 211
World's Fair, 1939 215–17, *203*
Wotton, Edward, *De differentiis animalium libri decem* 53
wraiths 97–8, 104

Wright, Judith, *The Coral Battleground* 261
Wroe, Ann, *Orpheus: The Song of Life* 79
Yanagawa, Shigenobu, *Yakuhinkai syuppinbutsu* 45, *45*
Yemen 157, 158, 159, 267, *144, Map 2*
Yonge, Charles Maurice 66, 129–30, 118
  A Year on the Great Barrier Reef 189
Yucatan (Mexico) 58, 241
Zanzibar 198, 267
Zheng He, 15th-century Chinese admiral and diplomat 159
Zoanthidea (order; zoanthid) 18, 183, 287, 290
  see also gold corals
Zocha welcome cup 162–3, *150*
Zola, Émile, *La Curée* 146
zombie ecosystems 219, 256
Zoophyta 8, 56, 290
zoophyte 7–10, 29, 32, 53, 56–7, 59, 66, 72, 91, 110–11, 139, 189, 192, 286, 290, *5, 104*
zooxanthellae
  clades of 41, 235
  and coral calcification 20
  discovery and naming of 65
  endosymbiosis with scleractinians 10, 12, 27, 29, 41–2, 66, 248, 285, 287, 288–90, *25, 39, 40*
  evolutionary origin of 40
  heat tolerance 236, 249
  light and oxygen-producing photosynthesis 24, 26, 41, 68, 248
  loss of in coral bleaching 72, 233–5, 286
  nutrient recycling with coral host 68, 72, 222
  resistance to photoinhibition 235
  see also dinoflagellates; *Symbiodinium*
Zoroaster 93, 157
Zucchi, Jacopo, *The Treasures of the Sea* (or The Allegory of the Discovery of the New World) 84, 85, 86, *76*
Zuni people 169–71, *157*